Handbook of Metal Forming Process

Handbook of Metal Forming Process

Edited by **Darren Wang**

New York

Published by NY Research Press,
23 West, 55th Street, Suite 816,
New York, NY 10019, USA
www.nyresearchpress.com

Handbook of Metal Forming Process
Edited by Darren Wang

International Standard Book Number: 978-1-63238-253-5 (Hardback)

Printed in the United States of America.

Contents

Preface

This book discusses various characteristics of metal forming and its process, tools and design. The various chapters within this book discuss advanced processes and analysis of these processes, keeping in mind the aspects of the materials. The book also includes chapters on machine tools and their structures. Strategies for a programmable metal forming press and procedures for calculating forming limits of sheet metal are also discussed.

The information shared in this book is based on empirical researches made by veterans in this field of study. The elaborative information provided in this book will help the readers further their scope of knowledge leading to advancements in this field.

Finally, I would like to thank my fellow researchers who gave constructive feedback and my family members who supported me at every step of my research.

Editor

Process

Hydroforming Process: Identification of the Material's Characteristics and Reliability Analysis

A. El Hami, B. Radi and A. Cherouat

Additional information is available at the end of the chapter

1. Introduction

The increasing application of hydroforming techniques in automotive and aerospace industries is due to its advantages over classical processes as stamping or welding. Particularly, tube hydroforming with various cross sectional shapes along the tube axis is a well-known and wide used technology for mass production, due to the improvement in computer controls and high pressure hydraulic systems (Asnafi et al., 2000; Hama et al, 2006; Cherouat et al., 2002). Many experimental studies of asymmetric hydroforming tube have been examined (Donald et al., 2000; Sokolowski et al., 2000). Theoretical models have been constructed to show the hydroforming limits, the material and the process parameters influence on the formability of the tube without failure (buckling and fracture) (Sokolowski et al.,2000). Due to the complexity of the process, theoretical studies up to date have produced relatively limited results corresponding the failure prediction. As for many other metal or sheet forming processes, the tendency of getting a more and more geometric complicated part demands a systematic numerical simulation of the hydroforming processes. This allows modifying virtually the process conditions in order to find the best process parameters for the final product. Thus, it gives an efficient way to reduce cost and time.

Many studies have been devoted to the mechanical and numerical modelling of the hydroforming processes using the finite element analysis (Hama et al., 2006; Donald et al., 2000), allowing the prediction of the material flow and the contact boundary evolution during the process. However, the main difficulty in many hydroforming processes is to find the convenient control of the evolution of the applied internal pressure and axial forces paths. This avoids the plastic flow localization leading to buckling or fracture of the tube during the process. In fact, when a metallic material is formed by such processes, it experiences large plastic deformations, leading to the formation of high strain localization zones and, consequently, to the onset of micro-defects or cracks. This damage initiation and its evolution cause the loss of the formed piece and indicate that the forming process itself should be modified to avoid the damage appearance (Cherouat et al.,2002). In principle, all materials and alloys used for deep drawing or stamping can be used for hydroforming applications as well (Koç et al,2002).

This chapter presents firstly a computational approach, based on a numerical and experimental methodology to adequately study and simulate the hydroforming formability of welded tube and sheet. The experimental study is dedicated to the identification of material parameters using an optimization algorithm known as the Nelder-Mead simplex (Radi et al.,2010) from the global measure of displacement and pressure expansion. Secondly, the reliability analysis of the hydroforming process of WT is presented and the numerical results are given to validate the adopted approach and to show the importance of this analysis.

2. Hydroforming process

For production of low-weight, high-energy absorbent, and cost-effective structural automotive components, hydroforming is now considered the only method in many cases.

The principle of tube hydroforming is shown in Figure 1. The hydroforming operation is either force-controlled (the axial forces vary with the internal pressure) or stroke-controlled (the strokes vary with the internal pressure). Note that the axial force and the stroke are strongly interrelated (see figure 1).

Force-controlled hydroforming is at the focus in (Asnafi et al.,2000), where the constructed analytical models are used to show

- which are the limits during hydroforming,
- how different material and process parameters influence the loading path and the forming result, and
- what an experimental investigation into hydroforming should focus on.

The hydroforming operation comprises two stages: free forming and calibration. The portion of the deformation in which the tube expands without tool contact, is called free forming. As soon as tool contact is established, the calibration starts.

Figure 1. The principle of tube hydroforming: (a) original tube shape and (b) final tube shape (before unloading).

During calibration, no additional material is fed into the expansion zone by the axis cylinders. The tube is forced to adopt the tool shape of the increasing internal pressure only.

Many studies have been devoted to the mechanical and numerical modeling of the hydroforming processes using the finite element analysis, allowing the prediction of the material flow and the contact boundary evolution during the process. However, the main difficulty in many hydroforming processes is to find the convenient control of the evolution of the applied internal pressure and axial forces paths. This avoids the plastic flow localization leading to buckling or fracture of the tube during the process. In fact, when a metallic material is formed by such processes, it experiences large plastic deformations, leading to the formation of high strain localization zones and, consequently, to the onset of micro-defects or cracks. This damage initiation and its evolution cause the loss of the formed piece and indicate that the forming process itself should be modified to avoid the damage appearance. In principle, all materials and alloys used for deep drawing or stamping can be used for hydroforming applications as well.

2.1. Mechanical characteristic of welded tube behaviour

Taking into account the ratio thickness/diameter of the tube, the radial stress is considerably small compared to the circumferential σ_θ and longitudinal stresses σ_z (see Figure 2). In addition, the principal axes of the stress tensor and the orthotropic axes are considered coaxial. The transverse anisotropy assumption represented through the yield criterion can be written as:

$$\bar{\sigma}^2 = F\left(\sigma_z - \sigma_\theta\right)^2 + G\sigma_z^2 + H\sigma_\theta^2 \tag{1}$$

with (F,G,H) are the parameters characterizing the current state of anisotropy.

If the circumferential direction is taken as a material reference, the anisotropy effect can be characterized by a single coefficient R and the equation (1) becomes:

$$\bar{\sigma}^2 = \frac{1}{1+R}\left[R\left(\sigma_z - \sigma_\theta\right)^2 + \sigma_z^2 + \sigma_\theta^2\right] \tag{2}$$

The assumptions of normality and consistency lead to the following equations:

$$\begin{cases} d\varepsilon_\theta = \dfrac{d\bar{\varepsilon}}{\bar{\sigma}}\left(\sigma_\theta - \dfrac{R}{1+R}\sigma_z\right) \\ d\varepsilon_z = \dfrac{d\bar{\varepsilon}}{\bar{\sigma}}\left(\sigma_z - \dfrac{R}{1+R}\sigma_\theta\right) \end{cases} \tag{3}$$

where $\bar{\varepsilon}$ is the effective plastic strain and $\left(\varepsilon_\theta, \varepsilon_z\right)$ are the strains in the circumferential and the axial directions. The effective strain for anisotropic material can be derived from equivalent plastic work definition, incompressibility condition, and the normality condition:

$$d\bar{\varepsilon} = \frac{\sqrt{1+R}}{\sqrt{1+2R}}\sqrt{d\varepsilon_z^2 + d\varepsilon_\theta^2 + R\left(d\varepsilon_z - d\varepsilon_\theta\right)^2} = \left(\sqrt{\gamma^2 + \frac{2R}{1+R}\gamma + 1}\right)\frac{1+R}{\sqrt{1+2R}}d\varepsilon_\theta \quad \text{with} \quad \gamma = \frac{d\varepsilon_z}{d\varepsilon_\theta} \tag{4}$$

Taking into account the relations expressing strain tensor increments, the equivalent stress (Equation 2) becomes:

$$\bar{\sigma} = \left(\sqrt{1+\gamma^2 + \frac{2R}{1+R}\gamma}\right)\frac{\sqrt{1+2R}}{1+R+R\gamma}\sigma_\theta \tag{5}$$

In the studied case, the tube ends are fixed. As a consequence, the longitudinal increment strain $d\varepsilon_z = 0$, and then relations (4) and (5) become:

$$\bar{\sigma} = \left(\sqrt{\frac{2R^2+3R+1}{(1+R)^3}}\right)\sigma_\theta \qquad\qquad d\bar{\varepsilon} = \left(\frac{1+R}{\sqrt{1+2R}}\right)d\varepsilon_\theta \tag{6}$$

The knowledge of the two unknown strain ε_θ and stress σ_θ needs the establishment of the final geometric data linked to the tube (diameter and wall thickness):

$$\varepsilon_\theta = \ln\left(\frac{d}{d_0}\right) \text{ and } \sigma_\theta = \frac{Pd}{2t} \tag{7}$$

where P is the internal pressure, (d, d_0) are the respective average values of the current and initial diameter of the sample and (t) is the current wall thickness obtained according to the following relation:

$$t = t_0 e^{-(1+\gamma)\varepsilon_\theta} \tag{8}$$

Finally, the material characteristics of the tube (base metal) are expressed by the effective stress and effective strain according to the following equation (Swift model):

$$\bar{\sigma} = K(\varepsilon_0 + \bar{\varepsilon})^n \tag{9}$$

The values of the strength coefficient K, the strain hardening exponent n, the initial strain ε_0 and the anisotropic coefficient R in Equations (2) and (9) are identified numerically. For the determination of the stress–strain relationship using bulge test, the radial displacement, the internal pressure and the thickness at the center of the tube are required.

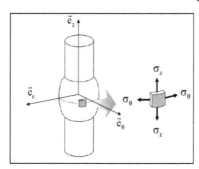

Figure 2. Stress state at bulge tip

3. Identification process

The parameters $\left(K, \varepsilon_0, n\right)$ are computed in such a way that the constitutive equations associated to the yield surface reproduce as well as possible the following characteristics of the sheet metal. The problem which remains to be solved consists in finding the best combination of the parameters damage which minimizes the difference between numerical forecasts and experimental results. This minimization related to the differences between the m experimental measurements of the tensions and their numerical forecasts conducted on tensile specimens.

Due to the complexity of the used formulas, we have developed a numerical minimization strategy based on the Nelder-Mead simplex method. The identification technique of the material parameters is based on the coupling between the Nelder-Mead simplex method (Matlab code) and the numerical simulation based finite element method via ABAQUS/Explicit© of the hydroforming process. To obtain information from the output file of the ABAQUS/Explicit©, we use a developed Python code (see Figure 3).

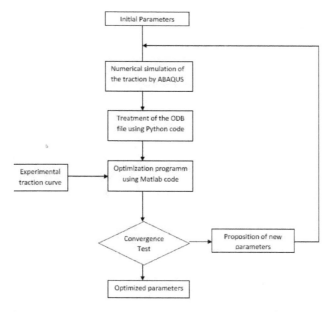

Figure 3. Identification process

4. Results and discussion

A three dimensional finite element analysis (FEA) has been performed using the finite element code ABAQUS/Explicit to investigate the hydroforming processes.

4.1. Tensile test

Rectangular specimens are made with the following geometric characteristics: thickness=1.0mm, width=12.52mm and initial length=100mm, were cut from stainless (Figure 4). All the numerical simulations were conducted under a controlled displacement condition with the constant velocity v=0.1mm/s. The predicted force versus displacement curves compared to the experimental results for the three studied orientations are shown in Figure 1. With small ductility (Step 1) the maximum stress is about 360MPa reached for 25% of plastic strain and the final fracture is obtained for 45% of plastic strain. With moderate ductility (Step 3) the maximum stress is about 394MPa reached in 37.2% of plastic strain and the final fracture is obtained for 53% of plastic strain. The best values of the material parameters using optimization procedure are summarized in Table 1. Within these coefficients the response (stress versus plastic strain) presents a non linear isotropic hardening with a maximum stress $\sigma_{max} = 279$ MPa reached in $\overline{\varepsilon}^P = 36.8\%$ of plastic strain and the final fracture is obtained for 22 % of plastic strain. The plastic strain map of the optimal case is presented in Figure 1.

Step	Critical plastic strain	K [MPa]	ε_0	n
1	25,8%	381,3	0.0100	0.2400
2	29,8%	395,5	0.0120	0.2415
3	37,2%	415,2	0.0150	0.2450
Optimal	36,8%	416,1	0.0198	0.2498

Table 1. Properties of the used material

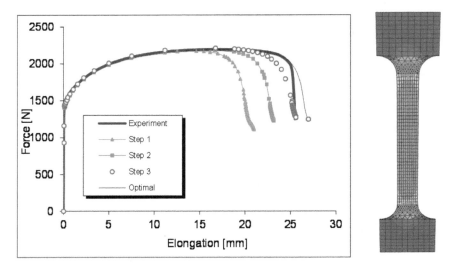

Figure 4. Force/elongation for different optimization steps and plastic strain map

4.2. Welded tube (WT) hydroforming process

In this case, the BM with geometrical singularities found in the WT is supposed orthotropic transverse, whereas its behaviour is represented by Swift model. The optical microscope observation on the cross section of the wall is used to build the geometrical profile of the notch generated by the welded junction. By considering the assumptions relating to an isotropic thin shell (R=1) with a uniform thickness, the previously established relations (6), (7) and (9) allow to build the first experimental hardening model using measurements of internal pressure/radial displacements. This model is then proposed, as initial solution, to solve the inverse problem of required hardening law that minimizes the following objective function:

$$\xi_F = \frac{1}{m_p} \sqrt{\sum_{i=1}^{m_p} \left(\frac{F_{exp}^i - F_{num}^i}{F_{exp}^i} \right)^2} \qquad (10)$$

where F_{exp}^i is the experimental value of the thrust force corresponding to i^{th} nanoindentation depth H_i, F_{num}^i is the corresponding simulated thrust force and m_p is the total number of experimental points.

Different flow stress evolutions of isotropic hardening (initial, intermediate and optimal) are proposed in order to estimate the best behavior of the BM with geometrical singularities found in the WT. Figures 5 and 6 show the effective stress versus plastic strain curves and the associate pressure/radial displacement for these three cases. As it can be seen, there is a good correlation between the optimal evolution of Swift hardening and the experimental results. Table 1 summarizes the parameters of these models.

Hardening model	ε_0	K (MPa)	n
Initial	0.025	1124.6	0.2941
Intermediate	0.055	692.30	0.2101
Optimal	0.080	742.50	0.2359

Table 2. Swift parameters of different hardening evolution

The anisotropy factor R is determined only for the optimal hardening evolution. In the problem to be solved there is only one parameter which initial solution exists, that it corresponds to the case of isotropic material (R = 1). The numerical iterations were performed on the WT with non-uniformity of the thickness (see Figure 7), and the obtained results are shown in Figure 8. A good improvement in the quality of predicted results is noted if R corresponds the value of 0.976.

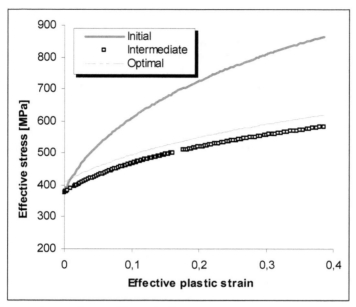

Figure 5. Stress-strain evolutions for different hardening laws

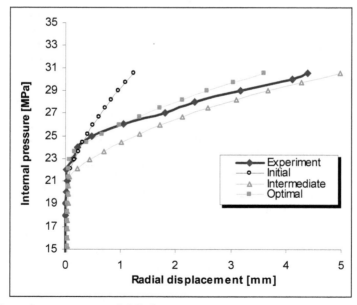

Figure 6. Internal pressure versus radial displacement

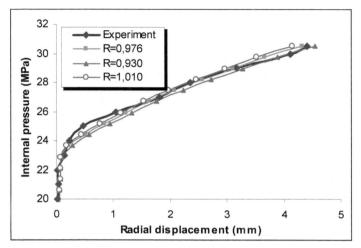

Figure 7. Radial displacement for different values of anisotropy coefficient R

4.3. Thin sheet hydroforming process

Sheet metal forming examples will be presented in order to test the capability of the proposed methodology to simulate thin sheet hydroforming operation using the fully isotropic model concerning elasticity and plasticity (Cherouat et al.,2008). These results are carried out on the circular part with a diameter of 300mm and thickness of 0.6mm. During hydroforming of the blank sheet, the die shape keeps touching the blank, which prevents the deformed area from further deformation and makes the deformation area move towards the outside. The blank flange is drawn into the female die, which abates thinning deformation of deformed area and aids the deformation of touching the female die and uniformity of deformation. Compared with the experiments done before, the limit drawing ratio of the blank is improved remarkably.

By considering the assumptions relating to an isotropic thin shell with a uniform thickness, the previously established relations allow to build the first experimental hardening model using measurements of force/displacement. This model is then proposed, as initial solution, to solve the inverse problem of required hardening law that minimizes the following objective function:

$$E_{error} = \frac{1}{m}\sqrt{\sum_{1}^{m}\left(\frac{P_{exp}^{i}-P_{num}^{i}}{P_{exp}^{i}}\right)^{2}} \qquad (11)$$

where P_{exp}^{i} is the experimental value of the thrust pressure corresponding to i^{th} displacement δ_{i}, P_{num}^{i} is the corresponding predicted pressure and m is the total number of experimental points.

The controlled process parameters are the internal fluid pressure applied to the sheet as a uniformly distributed load to the sheet inner surface and is introduced as a linearly increasing function of time with a constant flow from approximately 10 ml/min. The comparator is used to measure the pole displacement. The effect of three die cavities (D1, D2 and D3 see Figure 8) on the plastic flow and damage localisation is investigated during sheet hydroforming. These dies cavities are made of a succession of revolution surfaces (conical, planes, spherical concave and convex). The evolution of displacement to the poles according to the internal pressure during the forming test and sheet thicknesses are investigated experimentally. The profiles of displacements are obtained starting from the deformations of the sheet after bursting. Those are reconstituted using 3D scanner type Dr. Picza Roland of an accuracy of 5μm with a step of regulated touch to 5mm. In addition, two measurement techniques were used to evaluate the thinning of sheet after forming; namely a non-destructive technique using an ultrasonic source of Sofranel mark (Model 26MG) and a destructive technique using a digital micrometer calliper of Mitiyuta mark of precision 10μm (see Figure 9).

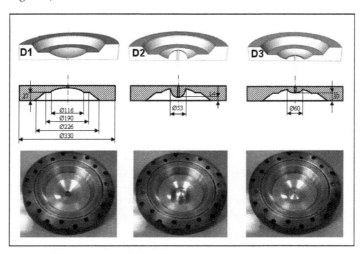

Figure 8. Geometry of die cavities (D1, D2 and D3)

Experiments results of circular sheet hydroforming are shown in Figure 10 (Die D1), Figure 11 (Die D2) and Figure 12 (Die D3). For the die cavities D1 and D3, fracture appeared at the round corner (near the border areas between the conical and the hemispherical surfaces of the die). For the die cavity D2, the fracture occurred at the centre of the blank when the pressure is excessive. This shows that the critical deformation occurs at these regions. It is noted that the rupture zone depends on the die overflow of the pressure medium from the pressurized chamber and the reverse-bending effect on the die shoulder were not observed in the experiment. In this part we are interested in the comparison between experimental observations of regions where damage occurred and numerical predictions of areas covered by plastic instability and damage. Figures 10, 11 and 12 present as such, the main results and simulations of all applications processed in this study. The predicted results with cavity dies show that the equivalent von Mises stress reached critical values high and then subjected to a significant decrease in damaged areas. This decrease is estimated for the three die cavities D1, D2 and D3, respectively 29%, 14% and 36%.

Figure 9. 3D scanner G Scan for reconstitution

Comparisons between numerical predictions of damaged areas and the experimental observations of fracture zones led us to the following findings:

1. The numerical calculations show that increasing pressure, the growing regions marked by a rise in the equivalent stress followed by a sudden decrease can be correlated with the damaged zones observed experimentally. In this context, the results of the first die cavity D1 show that instabilities are localized in the central zone of the blank, limited by a circular contour of the radius 72mm. The largest decrease in stress is located in the area bounded by two edges of respective radii 51 and 64mm. While the rupture occurred at the border on the flat surface with the spherical one located on a circle of radius 60mm. With the die cavity D2, the largest decrease is between two contours of radii 10 and 19mm, the rupture is observed at a distance of 17mm from the revolution axis of deformed blank. Finally with the die cavity D3, the calculations show that the damaged area is located in a region bounded by two edges of respective radii 54 and 73mm, the rupture occurred in the connection of the flat surface with the surface spherical one.
2. The pressures that characterize the early instabilities are respectively the order of 4.90MPa (for D1), 2.85MPa (for D2) and 5.1MPa (for D3). For applications with die cavities D1 and D3, regions where the beginnings of instability have been identified (see Table 2).

The results presented in Fig. 9 show that the relative differences between predicted and experiment results of pole displacement are in the limit of 7% while the pressure levels are below a threshold characterizing the type of application.

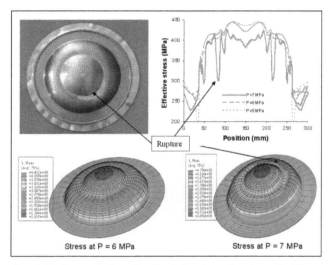

Figure 10. Experimental and numerical results of hydroforming using die cavity D1

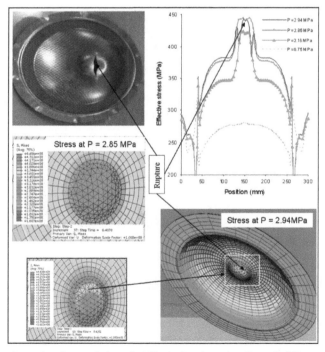

Figure 11. Experimental and numerical results of hydroforming using die cavity D2

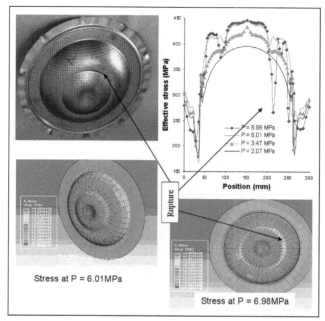

Figure 12. Experimental and numerical results of hydroforming using die cavity D3

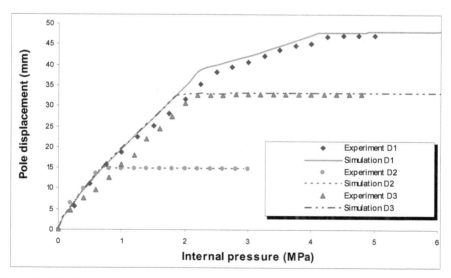

Figure 13. Pole displacement versus internal pressure

Die cavity	Beginning instability (MPa)	Critical (MPa)	Experimental (MPa)
D1	4,90	6,74	5,2
D2	2,85	2,85	3,0
D3	5,10	6,86	5,3

Table 3. Levels of pressure for different dies

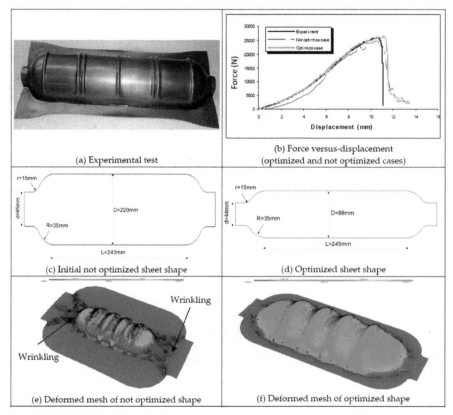

(a) Experimental test

(b) Force versus-displacement
(optimized and not optimized cases)

(c) Initial not optimized sheet shape

(d) Optimized sheet shape

(e) Deformed mesh of not optimized shape

(f) Deformed mesh of optimized shape

Figure 14. Optimisation of complex shape part

4.4. Optimization of sheet shape

Optimization is the action of obtaining the preferable results during the part design. In the CAE-based application of optimization, several situations can cause the numerical noise (wrinkling). When the numerical noise exists in the design analysis loop, it will create many artificial local minimums. In this case, the minimization of local thinning condition in the blank sheet metal was tested with a cost function of the optimization system was chosen to minimize the thinning ratio of 20% thinnest element.

i.e. Cost function: $f = \sum_{i=1}^{n} \left\| \frac{t-t_0}{t_0} \right\|^2$

where t_0 is the initial thickness and t the final thickness.

In this case a significant design variable for formability of blank during hydroforming process and the design (D and d) constraints were defined:

$50 \le D \le 250mm$ $20 \le d \le 100mm$. The experimental final shape is shown in Figure 10a. The comparison of the force versus the maximum displacement with the initial and optimized blank shape is present in Figure 10b. Good agreement between the optimum shape and the experimental values. Figures 10c and 10d compare the initial and the optimum blank shape. Successfully decreased the cost function (thinning ratio) from 50% to 20% is obtained without wrinkling (Figure 10e and 10f) (see Ayadi et al.,2011).

5. Reliability analysis

Recently, RBDO has become a popular philosophy to solve different kind of problem. In this part, we try to prove the ability of this strategy to optimize loading path in the case of THP where different kind of nonlinearities exist (material, geometries and boundary conditions). The aim of this study is to obtain a free defects part with a good thickness distribution, decrease the risk of potential failures and to let the process insensitive to the input parameters variations. For more detailed description of the RBDO methodology and variety of frameworks the reader can be refer to the following references (Youn et al., 2003; Enevoldsen et al., 1994; El Hami et al., 2011). The RBDO problem can be generally formulated as:

$$\begin{cases} \text{Min } f(d,X) \\ \text{subject to } P\left[G_i(d,X) \le 0\right] - \Phi\left(-\beta_{t_i}\right) \le 0 \; i = 1,...,np \\ d^L \le d \le d^U, \; d \in R^{ndv} \text{ and } X \in R^{nrv} \end{cases} \tag{12}$$

where f(d,X) is the objective function, d is the design vector, X is the random vector, and the probabilistic constraints is described by the performance function $G_i(X)$, np, ndv and nrv are the number of probabilistic constraints, design variables and random variables, respectively, β_{ti} is the prescribed confidence level which can be defined as $\beta_{t_i} = -\Phi^{-1}(P_f)$ where P_f is the probability of failure and Φ is the cumulative distribution function for standard normal distribution.

The process failure state is characterized by a limit state function or performance function G(X), and G(X)=0 denotes the limit state surface. The m-dimensional uncertainty space in thus divided into a safe region $\left(\Omega_s = \{X : G(X) > 0\}\right)$ and a failure region $\left(\Omega_f = \{X : G(X) \le 0\}\right)$ (see Radi et al.,2007).

5.1. Definition of the limit state functions

The risk of failure is estimated based on the identification of the most critical element for necking and severe thinning. For this reason fine mesh was used in this study to localize the plastic instability or the failure modes in one element. Some deterministic finite element simulations show that always severe thinning is localised in element 939 in the centre of the expansion region and necking in element 1288 as shown in Figure 22.

Since the strain\stress of element 939\1288 is the critical strain\stress of the hydroformed tube, then the reliability of these two elements represented in reality the reliability of the hydroforming process.

In this work, the limit state functions take advantage from the FLSD and the FLD of the material to assess the risk or the probability of failure of necking and severe thinning. From these curves we distinguish mainly two zones: feasible region: when a tube hydroforming process can be done in secure conditions and unfeasible region when plastic instability can appear as shown in Figures 24-25. In reality, the FLSD and FLD was used in several papers (Kleiber et al, 2004; Bing et al., 2007) as failure criteria in the aim to assess the probability of failure.

The limit state function depends on the variable of the process. Mathematically, this function can be described as $Z = G(\{x\}, \{y\})$, where $\{x\}$ presents a vector of deterministic variables and $\{y\}$ is a vector of random variables.

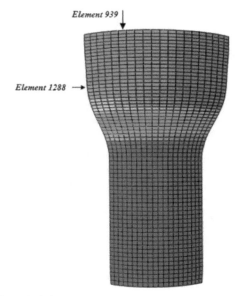

Figure 15. Location of the critical elements for severe thinning and necking

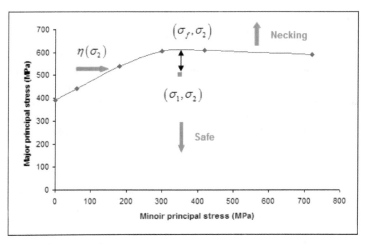

Figure 16. Forming limit stress diagram

The first limit state function was taken to be the difference between the maximum stress and the corresponding FLSD as shown in Figure 25:

$$\begin{cases} G(\{x\},\{y\}) = \sigma_f - \sigma_1^c \\ \sigma_f = \eta(\sigma_2) \end{cases} \tag{13}$$

where σ_1^c is the maximum stress in the most critical element and σ_f the corresponding forming stress limit. The role of this constraint is to maintain the maximum stress on the critical element below σ_f. The second limit state function is used to control the severe thinning in the tube, to define this function we use the FLD plotted in the strain diagram as shown in Figure 25, it can be given by the following expression:

$$\begin{cases} G(\{x\},\{y\}) = F_{th} = \sigma_f - \sigma_1^c \\ \sigma_f = \eta(\sigma_2) \end{cases} \tag{14}$$

where ε_1 is the major strain in the critical element and ε_{th} is the thinning limit determined from the FLD curve as shown in Figure 17.

The objective function consists to reduce the wrinkling tendency, this function is inspired from the FLD and given by the following expression:

$$\begin{cases} F_w = \sum_{i=1}^N (d_w^i)^2 = \sum_{i=1}^N (\varepsilon_1^i - \varepsilon_w^i)^2 \\ \varepsilon_w = \phi(\varepsilon_2) \end{cases} \tag{15}$$

where ε_1 is the major strain in element i, and ε_w is the wrinkling limit value determined from the FLD, N is the number of elements.

The success of a THP is dependent on a number of variables such as the loading paths (internal pressure versus time and axial displacement versus time), lubrication condition, and material formability. A suitable combination between all these parameters is important to avoid part failure due to wrinkling, severe thinning or necking. Koç et al. (Koç et al.,2002) found that loading path and variation in material properties has a significant effect on the robustness of the THP and final part specifications. In this work, we define the load path as design variables to be optimized.

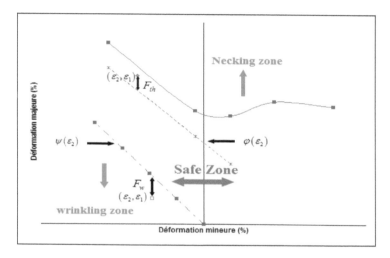

Figure 17. Forming Limit Diagram

The load path given the variation of the inner pressure vs. time is modelled by two points (P_1, P_2) displacement is imposed as a linear function of time, for axial displacement we interest only on the amplitude D. Table 4 illustrates the statistical properties of the design variables.

Variable	Mean value	Cov(%)	Distribution type
P_1 (MPa)	15	5	Normal
P_2 (MPa)	35	5	Normal
D(mm)	8	5	Normal

Table 4. Statistical properties of the control points described the load path

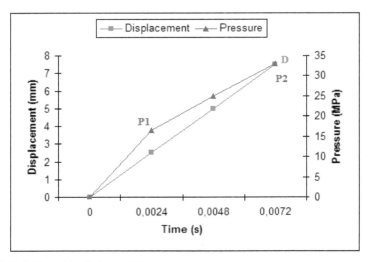

Figure 18. Definition of the design variables

5.2. Definition of the random variables

In real metal forming processes the material properties of the blank may vary within a specific range and thus probably also impact the forming results. In this work, the material of the tube is assumed to be isotropic elastic-plastic steel obeying the power-law:

$$\sigma = K(\varepsilon + \varepsilon_0)^n \tag{16}$$

where K is the strength coefficient value, n the work hardening exponent, ε_0 the strain parameter, and ε the true strain. Hardening variables (K, n) are assumed to be normal distributed with mean values μ and standard deviations σ. Friction problem plays also a key role in hydroforming process and present some scatter, to take account for this variation a normal distribution of the static friction coefficient is assumed. Finally, the initial thickness of the tube is considered as a random variable. Table 5 illustrates the statistical properties of all random parameters.

Variable	Mean value	Cov(%)	Distribution type
K(MPa)	530	5	Normal
n	0.22	5	Normal
h(mm)	1	5	Normal
μ	0.1	5	Normal

Table 5. Statistical properties of random parameters

We make the assumption that all the input parameters are considered to be statistically independent.

5.3. Evaluation of the probability of failure

Consider a total number of m stochastic variables denoted by a vector $X = \{x_1, x_2, \ldots, x_m\}^T$, in probabilistic reliability theory, the failure probability of the process is expressed as the multi-variant integral:

$$P_f = P\big(G_j(x) < 0\big) = \int_{\Omega_j} f_X(x)dx_1 \ldots dx_n \quad \text{where } \Omega_j : \big\{x \in \Re^n : G_j(x) < 0\big\} \tag{17}$$

where P_f is the process failure probability, $f_X(X)$ is the joint probability density function of the random variables X. A reliability analysis method was generally employed since been very difficult to directly evaluate the integration in Equation (20). In the case when the problem presents a high non linearity, the use of the classical method to assess the probability of failure becomes impracticable.

Evaluation of the probability of failure is metal forming processes remain still a complicated and computational cost due to the lot of parameters that can be certain and the absence of an explicit limit state function. The appliance of the direct Monte Carlo seems impractical.

Therefore various numerical techniques have been proposed for reducing the computational cost in the evaluation of the probability of failure (Donglai et al., 2008; Jansson et al., 2007). Monte Carlo simulations coupled with response surface methodology (RSM) is used to assess the probability of failure. To build the objective function and the limit state functions given by Equations (16), (17) and (18), RSM is used based on the use Latin Hypercube design (LHD). The LHD was introduced in the present work for its efficiency, with this technique; the design space for each factor is uniformly divided. These levels are the randomly combined to specify n points defining the design matrix. Totally 50 deterministic finite element simulations were run, from these results we find a suitable approximation for the true functional relationship between response of interest y and a set of controllable variables that represent the design and random variables. Usually when the response function is not known or non Linear, a second order is utilized in the form:

$$y = \beta_0 + \sum_{i=1}^{n}\beta_i x_i + \sum_{i=1}^{n}\beta_{ii}^2 x_i^2 + \sum_{i<j}^{n}\sum_{j=1}^{n}\beta_{ij}x_i x_j + \varepsilon \tag{18}$$

where ε represents the noise or error observed in the response, y such that the expected response is $y - \varepsilon$ and β's the regression coefficients to be estimated. The least square technique is being used to fit a model equation containing the input variables by minimizing the residual error measured by the sum of square deviations. To assess the probability of failure, the limit state functions are then estimated for a new more consequent sample (1 million) starting from the model given by the response surface methodology and the probability of failure is given then by $P_f = N_{fail}/N_{total}$. N_{fail} is the number of failing points and N_{total} is the total number of simulations. This methodology will be implemented in the optimization process to optimize the loading paths with taking into account of the uncertainty associated to the parameters defined previously. The method presented here seems more suitable since optimization of the metal forming processes is time consuming and require many evaluations of the probabilistic constraints, additional it can be used in conjunction with an optimization procedure.

In order to verify the quality of the response surface, a classical measure of the correlation between the approximate models and the exact value given by finite element simulations of the limit state function is used and shows that the approximation models can predict with a high precision the real response. Before proceeding to the reliability analysis and optimization process, the main effect plot is drawn to show how each of the variables affect severe thinning and necking. It is observed that the strength coefficient, work hardening exponent and initial thickness of the tube has the most significant impact on the severe thinning and necking plastic instabilities.

5.4. Finite element model

Figure 28 shows a finite element model (FEM) that was defined to simulate the THP. It is formed of the die that represents the final desired part, a punch modelled as rigid body and meshed with 4-node, bilinear quadrilateral, rigid element called R3D4. The tube is composed of 1340 elements (4-node, reduced integration, doubly curved shell element with five integration points through the shell section, called S4R). Since the part is symmetrical, the only quarter-model was used. The numerical simulations of the process are carried out using the explicit dynamic finite element code ABAQUS\Explicit©. The dynamic explicit algorithm seems more suitable for this simulation.

5.5. Formulation of the optimization problem

In this work, we aim to optimize the loading path under the variation of some parameters, here the objective function consist to minimize the wrinkling tendency and the probabilistic constraints was defined to avoid severe thinning and necking. We can formulate the RBDO problem as follows:

$$\begin{cases} \text{Min } f(p) \\ \text{s.c to } P\Big[G_{thinning}(p,X) \le 0\Big] \le Pa_i \\ \qquad P\Big[G_{necking}(p,X) \le 0\Big] \le Pa_i \end{cases} \qquad (19)$$

where $P\Big[G_i(p,X) \le 0\Big]$ and Pa_i are the probability of constraint violation and the allowable probability of the ith constraint violation, respectively.

A probabilistic methodology was developed and applied to optimize THP with respect to probabilistic constraint. The methodology combined an optimization strategy and probabilistic analysis. A routine is prepared with MATLAB with the use of the toolbox optimization strategy based a successive quadratic programming. The probabilistic methodology allow to take account to the variability in metal forming process particularly is known that theses uncertainty have a significant impact on the success or the failure of the process and the quality of the final part.

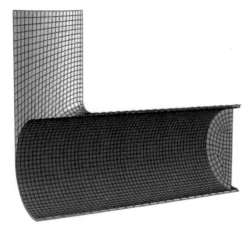

Figure 19. Finite Element Model

In general manner the RBDO is solved in two spaces physical space for the design variables and normal space when we assess the reliability index. In order to avoid calculation of the reliability and the separation of the solution in two spaces which leads to very large computational time especially for large scale structures and for high nonlinear problem like hydroforming process, the transformation approach that consist in finding in one step the probability of failure based on the predicted models and optimal design is used. In this methodology, a deterministic optimization and a reliability analysis are performed sequentially, and the procedure is repeated until desired convergence is achieved.

5.6. Results and discussion

Optimization problem is solved with different reliability level target or allowable probability of failure: $P_f = 2.28\% \Leftrightarrow \beta = 2; P_f = 0.62\% \Leftrightarrow \beta = 2.5; P_f = 0.13\% \Leftrightarrow \beta = 3$.

Table 6 resume the results obtained in the case of deterministic and reliability design for different values of the reliability index. The resolution of the problem shows that the deterministic design presents a high probability of failure for necking but an acceptable probability of failure in severe thinning, the benefits of RBDO is to ensure a level of reliability for both necking and severe thinning. The results of the optimization are reported in Table 6.

where β_1 is the reliability level for necking and β_2 for severe thinning. As shown is table 3, for the deterministic design we have a high reliability level for severe thinning compared to the reliable design but this, it's not true for necking, in fact optimization based on reliability analysis try to find a tradeoffs between the desired reliability confidence.

Design	$D(mm)$	$P_1(MPa)$	$P_2(MPa)$	β_1	β_2
DDO	7	18	35.4643	1.6418	4.6112
RBDO $\beta = 2$	7	17.1111	35	1.9995	4.0376
RBDO $\beta = 2.5$	7	16.1165	35	2.5004	3.5679
RBDO $\beta = 3$	7.1132	14.9451	35.0035	3.0005	3.1669

Table 6. Optimal parameters for different design

The main drawback of RBDO is that it requires high number of iterations compared to deterministic approach to converge. Table 7 shows the percentage decrease of the objective function and the iterations number for the different cases.

Design	Deterministic	$\beta = 2$	$\beta = 2.5$	$\beta = 3$
% of decrease	35.7754	33.7909	30.0629	23.4896
SQP iteration	10	19	21	19

Table 7. Decrease of the objective function and number of iterations

Figure 20 presents the thickness distribution in an axial position obtained with deterministic approach and for the optimization strategy with the consideration to the probabilistic constraints. With a probabilistic approach satisfactory results are obtained to achieve a better thickness distribution in the tube (El Hami et al.,2012).

To show the effects of the introduced variability on the probabilistic constraints, a probabilistic characterization of severe thinning and necking when β=0 has been carried out. The generalized extreme value distributions type I (k=0) for severe thinning and type III (k<0)for necking seem fit very well the data. The probability density function for the generalized extreme value distribution with location parameter ⊕, scale parameter σ, and shape parameter $k \neq 0$ is:

$$f(x|k,\mu,\sigma) = \left(\frac{1}{\sigma}\right)\exp\left(-\left(1+k\frac{(x-\mu)}{\sigma}\right)^{-\frac{1}{k}}\right)\left(1+k\frac{(x-\mu)}{\sigma}\right)^{-1-\frac{1}{k}} \tag{20}$$

For $k = 0$, corresponding to the Type I case, the density is:

$$f(x|0,\mu,\sigma) = \left(\frac{1}{\sigma}\right)\exp\left(-\exp\left(-\frac{(x-\mu)}{\sigma}\right) - \frac{(x-\mu)}{\sigma}\right) \tag{21}$$

The parameters that characterize these distributions are summarized in Table 8. Then we can simply assess the probability of failure of the potential failure modes to show how uncertainties can affect the probability of failure.

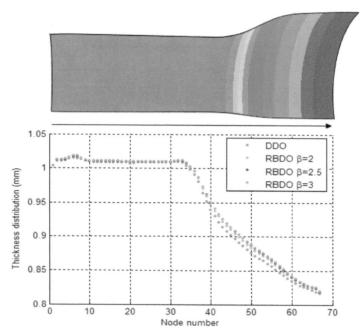

Figure 20. Thickness variation in an axial position

Parameters	μ	σ	k
$G_{necking}$	264.8466	54.1678	0
$G_{thinning}$	0.2223	0.0882	−0.0654

Table 8. Statistical parameters of the extreme value distribution

6. Conclusion

The first part of this work presents the results of a combined experimental/numerical effort that aims to assess the performance of different plastic stress flow in predicting the burst of welded steel tubes loaded under internal pressure. The prediction of the stress-strain characteristic with the anisotropic effect of tubular material has been proposed. Once the expanding diameter, the internal pressure and the wall thickness were obtained from the results of the bulge tests, the effective stress and effective strain could be calculated. The bulge tests carried out until bursting showed that all the fissures are initiated in the central area of the expanding zone not far from the weld zone.

Using the Nelder-Mead (NM) simplex search method, a flow stress curve (Swift's model) that best fits the stress-strain of the used anisotropic material could be determined with consideration global response (force/displacement). The local behaviour (stress/strain) of the welded joints and the HAZ is identified numerically using ABAQUS solver from global results (force/depth) of nanoindentation tests. The identified hardening coefficients are introduced by Swift model. From the simulations carried out, it is clear the influence of the plastic flow behaviour of the WT in the final results (thickness distribution, stress instability, tube circularity and critical thinning and rupture).

It is also clear that to predict with more accuracy the results, the model used for simulation has to be as realistic as possible. Therefore, future work in this area will include the experimental identification approach of the hardening model coupled with damage. Indeed, we think that measurements of displacements and strains without contact can improve results quality. The suggested model coupled with ductile damage can contribute to the deduction of forming limit diagrams.

The plastic deformation of a circular sheet hydraulically expanded into a complex female die was explored using experimental procedure and numerical method using ABAQUS/EXPLICIT code©. As future work, one can study others optimization techniques without using derivatives to make a numerical comparison between these different techniques and integration of adaptive remeshing procedure of sheet forming processes.

In the second part of this work, an efficient method was proposed to optimize the THP with taking into account the uncertainties that can affect the process. The optimization process consists to minimize an objective function based on the wrinkling tendency of the tube under probabilistic constraints that ensure to decrease the risk of potential failure as necking and severe thinning. This method can ensure a stable process by determining a load path that can be insensitive to the variations that can affect input parameters. Construction of the objective function and reliability analysis was done based on the response surface method (RSM). The study shows that the RSM is an effective way to reduce the number of simulations and keep a good accuracy for the optimization.

Probabilistic approach revealed several advantages and promoter way than conventional deterministic methodologies, however, probabilistic approach need precise information on the probability distributions of the uncertainty and is sometimes scarce or even absent. Moreover, some uncertainties are not random in nature and cannot be defined in a probabilistic framework.

Author details

A. El Hami
LMR, INSA de Rouen, St Etienne de Rouvray, France

B. Radi
LMMI, FST Settat, Settat, Morocco

A. Cherouat
GAMMA3, UTT, Troyes, France

7. References

Asnafi, N. & Skogsgardh, A. (2000). Theoretical and experimental analysis of stroke-controlled tube hydroforming, *Materials Science and Engineering A279*, pp. 95-110

Ayadi, A., Radi, B., Cherouat, A. & El Hami A. (2011). Optimization and identification of the characteristics of an Hydroformed Structures, *Applied Mechanics and Materials*, pp. 11-20.

Bing, L. et al. (2007). Improving the Reliability of the Tube-Hydroforming Process by the Taguchi Method, *Journal of Pressure Vessel Technology*, 129, pp. 242-247.

Cherouat, A., Saanouni, K. & Hammi, Y. (2002). Numerical improvement of thin tubes hydroforming with respect to ductile damage, *Int. J. of Mech. Sciences* 44, pp.2427-2446

Cherouat, A., Radi, B. & El Hami, A. (2008). The frictional contact of the composite fabric's shaping, *Acta Mechanica*, DOI 10.1007/s00707-007-0566-1.

Donald, B. J. & Hashmi, M.S.J. (2000). Finite element simulation of bulge forming of a cross-joint from a tubular blank, *J. of Materials Processing Technology* 103, pp. 333-342.

Donglai, W. et al. (2008). Optimization and tolerance prediction of sheet metal forming process using response surface model, *Computational Materials Science*, 42, pp. 228-233.

El Hami, A. & Radi, B. (2011). Comparison Study of Different Reliability-Based Design Optimization Approaches, *Advanced Materials Research*, pp. 119-130.

Enevoldsen, I. & Sorensen, JD. (1994). Reliability-based optimization in structural engineering, *Struct Safety*, 15, pp. 169–96.

Hama, T., Ohkubo, T., Kurisu, K., Fujimoto, H. and Takuda, H. (2006). Formability of tube hydroforming under various loading paths, *J. of Materials Processing Technology*, 177, pp. 676-679.

Jansson, T. et al. (2007). Reliability analysis of a sheet metal forming process using Monte-Carlo analysis and metamodels, *Journal of Materials Processing Technology*, 202, pp. 255-268.

Kleiber, M. et al. (2004). Response surface method for probabilistic assessment of metal forming failures, *International Journal for Numerical Methods in Engineering*, 60, pp. 51-67.

Koç, M. et al. (2002). Prediction of forming limits and parameter in the tube hydroforming process, *International Journal of Machine Tools and Manufacture*, 42, pp. 123-138.

Radi, B., Cherouat, A., Ayadi, M. & El Hami, A. (2010). Materials characterization of an hydroformed structure, *International Journal Simulation of Multidisciplinary Design Optimization*, 4, pp. 39-47.

Radi, B., El Hami, A. & Cherouat, A. (2012). Reliability Based Design Optimization Analysis of Tube Hydroforming Process, *International Journal Simulation*, in press.

Radi, B. & El Hami, A. (2007). Reliability analysis of the metal forming process, *Mathematical and Computer Modelling*, 45, pp. 431-439.

Sokolowski, T., Gerke, K., Ahmetoglu, M. & Atlan, T. (2000). Evaluation of tube formability and material characteristics: hydraulic bulge testing of tubes, *Journal of Materials Processing Technology* 98, pp. 34-40.

Youn, Byeng D, et al. (2003). A new response surface methodology for reliability-based design optimization, *Computers and structures*, 82, pp. 241-256.

Chapter 2

Stamping-Forging Processing of Sheet Metal Parts

Xin-Yun Wang, Jun-song Jin, Lei Deng and Qiu Zheng

Additional information is available at the end of the chapter

1. Introduction

1.1. Principle of stamping-forging processing (SFP) for sheet metal

SFP is a combined metal forming technology of stamping and forging for sheet metal parts. In an SFP, generally, stamping or drawing is used to form the spatial shape of the part first, and followed by a bulk forming employed to form the local thickened feature. It is suitable for making sheet metal parts which have local thickened feature, such as single or double layers cup parts with thickened inner or outer wall, disc-like parts with thickened rim, etc.

1. SFP principle of thickening in axial direction

It is difficult to manufacture cup part whose wall thickness is greater than that of its bottom by general sheet metal forming technology. As is well known, on the one hand, making the thickness of sheet metal reduction by compression is almost impossible due to the great metal flow resistance. On the other hand, it is not able to obtain different thicknesses of wall and bottom by stamping, although it is very effective to form the spatial shape of sheet metal part. The SFP of thickening in axial direction is just feasible to manufacture this type part.

The SFP principle of forming this kind of parts is shown in Fig .1. Firstly, a disk blank is formed to a single or double layer cylinder cup by a conventional drawing or hole flanging process. And then the inner or outer wall is thickened by an axial upsetting process. In the thickening step, the non-freestyle upsetting process such as hydraulic pressure assistant upsetting and small gap upsetting with rigid support will be used. In the former upsetting, a hydraulic pressure is adopted to make the blank stick with the sidewall of die to ensure stability (see Fig.1 a). In the latter upsetting, a mandrel is placed in the center hole. When the wall comes with a slight local thickening, the other cavity wall will contact it immediately to stop a further instability causing folding defect.

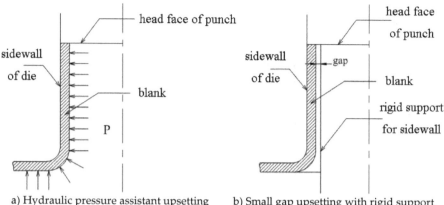

a) Hydraulic pressure assistant upsetting   b) Small gap upsetting with rigid support

Figure 1. Diagram of axial thickening SFP

In the hydraulic pressure assistant upsetting, the sidewall will be thickened greatly because the stabilities of sidewall of the blank can be guaranteed effectively by the assist of hydraulic pressure. The die for this upsetting process is complicated and its application is greatly restricted by high sealing requirements of the whole structure. Whereas, the instability of the parts wall in a small gap upsetting with rigid support can be controlled by increasing the upsetting step or decreasing the amount of thickening. Compared with the hydraulic pressure assistant upsetting, the dies of small gap upsetting with rigid support are simpler, in which, only a mandrel is needed to put in the center hole of part, and the gap between punch and mandrel could be changed by changing the diameter of the tooling.

2.   SFP principle of radial thickening process

The principle of radial thickening process is shown in Fig. 2 and Fig. 3. Firstly, a conventional drawing is used to form a lower boss in the center of the preformed part (Fig. 2). Then the preformed part is clamped in the center by the upper spindle and lower spindle of spinning machine (Fig. 3). When the preformed part rotates together with the spindles, the rollers feed in radial direction and thicken the rim one by one. The lower mandrel inserted into lower spindle can stop the part sliding in the radial direction between the spindles when bears asymmetric radial force during spinning.

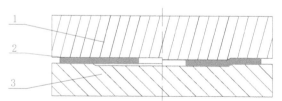

1- Upper die 2-Blank 3-Lower die

Figure 2. Perform by stamping

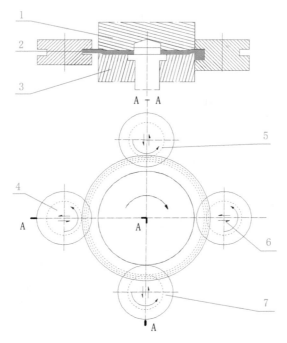

1-Upper spindle 2-Preformed part 3-Lower spindle 4,5,6,7-Rollers

Figure 3. Radial thickening by spinning

1.2. Classification of SFP

In terms of the combination mode of stamping and forging, the SFP process can be classified as compound SFP and sequence SFP. In the compound SFP, the stamping process and local bulk forming are carried out in one die-set. In the sequence SFP, the stamping process and subsequent local bulk forming are carried out in different die-sets.

In terms of the tool movement, the SFP can be also classified as linear SFP and rotational SFP. In the linear SFP, the tools move along a line and thicken the local feature wholly in one or more axial upsetting steps. In the rotational SFP, the tools feed along a radial direction and thicken the rim of the blank incrementally.

1.3. State-of-the-art of SFP

With the increasing demand of lightweight and high properties of parts, SFP has become a research hotspot in the field of metal forming [1]. More and more parts were made by SFP instead of conventional method [2]. Tan et al. developed a two-stage forming process of tailor blanks having local thickening for controlling the distribution of wall thickness of stamping parts. In the first stage, the target portion of the sheet for the local thickening was drawn into the die cavity, and then the bulging ring was compressed with the flat die under the clamping of the flange portion in the second stage [3]. Mori et al. developed a two-stage cold stamping process for forming magnesium alloy cups having a small corner radius from commercial magnesium alloy sheet. In the first stage, a cup having large corner radius was formed by deep drawing using a punch having large corner radius, and then the corner radius of the cup was decreased by compressing the side wall in the second stage. In the deep drawing of the first stage, fracture was prevented by decreasing the concentration of deformation with the punch having large corner radius. The radii of the bottom and side corners of the square cup were reduced by a rubber punch for applying pressure at these corners in the second stage [4]. Mori et al. also developed a plate forging process of tailored blanks having local thickening for the deep drawing of square cups to improve the drawability. A sheet having uniform thickness was bent into a hat shape of two inclined portions, and then was compressed with a flat die under restraint of both edges to thicken the two inclined portions. The bending and compression were repeated after a right-angled rotation of the sheet for thickening in the perpendicular direction. The thickness of the rectangular ring portion equivalent to the bottom corner of the square cup was increased, particularly the thickening at the four corners of the rectangular ring undergoing large decrease in wall thickness in the deep drawing of square cups became double [5]. Wang et al. prompted a drawing-thickening technology with axial force for double-cup shape workpieces by combining the characteristics of cold extrusion with drawing process [6]. An axial thrust was exerted to the sidewall during backward drawing to achieve the purpose of drawing and thickening [7]. Wang et al. also adopted SFP to form a flywheel plate and a sleeve with thickened wall instead of a traditional process, such as cutting and weld assembling [8,9].

Compared with traditional metal forming methods joining parts of different thickness by welding, the SFP method mentioned above can not only decrease the cost, but also can produce high quality sheet metal parts with shorten supply chains. With the development of industry, especially automotive industry, large quantities of parts with different wall thickness are needed. Thus, it is important to research SFP technology to manufacture such kind of sheet metal parts.

2. Thickening of outer wall of cup parts with axial upsetting

2.1. Design of thickening process and thickening ratio in single upsetting

In the SFP of cup parts with thickened wall, the axial upsetting of the wall is similar to tube upsetting. There are four situations of the forged piece formed from tube stock by upsetting processing: inner diameter remained and outer diameter enhanced, inner diameter decreased and outer diameter remained, inner diameter decreased and outer diameter enhanced, both inner and outer diameter enhanced and the thickness of the wall of tube is unchanged simultaneously. For the sheet metal upsetting thickening processing, there is no deformation mode that both inner and outer diameter enhanced and thickness is basically unchanged. In this section we mainly talk about the sheet metal upsetting thickening processing that inner diameter decreased and outer diameter remained.

For the cup part with thickened outer wall, the axial upsetting can be used to thicken the wall after drawing. The schematic of outer wall thickening process is shown in Fig. 4. At first stage, the sheet metal with uniform thickness is drawn into the die cavity by large round corner punch for preventing the occurrence of fracture. At second stage, the formed cup is ironed firstly by small round corner punch to make bottom to specified dimension. Then, a circular upsetting punch compress the outer wall to a thickened dimension and make the outer round corner to a specified radius.

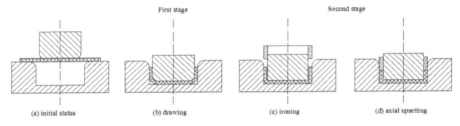

Figure 4. Schematic of outer wall thickening process.

Due to the wrinkling is easy to occur during axial upsetting, it's important to determine the limitation of thickening. In this section, thickening ratio of upset thickness to initial thickness is presented to define the formability. There are several geometry parameters play important roles in thickening ratio, such as wall height and inner corner radius, etc. The allowable thickening ratio under different conditions is shown in Fig. 5. The digits in the

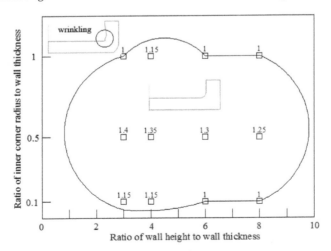

Figure 5. Allowable thickening ratio under different conditions.

figure are the thickening ratios obtained from simulation results in which the outer diameter of part is 120 mm. The number 1 represents the part occurred folding under corresponding conditions. It can be seen that the zone enclosed by lines is suitable for thickening the outer wall. When the ratio of inner corner radius to wall thickness is about 0.5, the thickening ratio has the largest value, which achieves to 1.4. With the increasing of wall height, the thickening ratio decreases under the condition of any inner corner radius.

2.2. Calculation of forming load

2.2.1. Upsetting force F_u

The axial upsetting is a closed-die forging process. Although the metal flow is different from that of tubing upsetting at beginning, the final upsetting force is similar to tubing upsetting force. So, according to calculation method of tubing upsetting force, the upsetting force can be expressed as an empirical formula:

$$F_u = 1.3pA$$

where p is the average upsetting stress (MPa), A is the area of wall section (mm^2).

2.2.2. Methods for reducing forming load

Because the axial upsetting is a closed-die forging process, the upsetting force increases rapidly at the end of metal forming. Relief cavity can be set in dies or blank to increase free flow surface and avoid full closing realizing reduction of upsetting force. The design of relief cavity can adopt the following three styles: a) center hole relief at the bottom of cup, b) relief cavity at the dies corresponding to outer corner between wall and bottom, c) combined relief style.

1. Hole relief

    If there is a center hole at the bottom of cup, hole relief method can be used to reduce upsetting force. In this method, piercing process must be carried out before axial upsetting. Then, while the outer wall is upset axially, the metal is enforced to flow to the center of the bottom, at where there is a free flow surface. The metal forming is no longer a closed die forging. So, the upsetting force can be decreased. Certainly, as a result of axial upsetting, the diameter of hole is decreased during the wall thickening. It is suggested that the center hole of part can be obtained by means of designing a proper relief hole which will be shrunk to specified dimension. This is able to reduce the upsetting force as well as avoid piercing again.

2. Relief cavity

    From the metal flow of upsetting process mentioned in above section, outer corner between wall and bottom is the last formed position of part. During forming this position, the upsetting force increases rapidly. It can be considered to adding a relief cavity in the die corresponding outer corner, which can increase free flow surface and decrease the upsetting force. After upsetting, machining should be carried out to clear the unnecessary metal away.

The position and shape of relief cavity can be designed as two modes as shown in Fig. 6: a) at the bearing plate under the bottom of the cup, which needs to manufacture a circular cavity in bearing plate; b) at the die and bearing plate, which needs to manufacture circular cavities in bearing plate and cylinder die respectively.

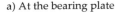

a) At the bearing plate   b) At the die and bearing plate

1-Bearing plate, 2-Cylinder die

Figure 6. Mode of relief cavity.

As using relief cavity method, the metal flow is similar to that of without relief cavity before the die fully filled. There is just a few of material enforced to flow into relief cavity after the wall formed. Finally, the cavity is not fully filled, which remain a little of free surface resulting in a decrease in upsetting force.

3.   Combined relief method

The two relief methods mentioned above can be used together. A center hole is pierced before upsetting, as well as a relief cavity is designed in the dies. In this way, the upsetting force relief can be more than that of single method, just not significant.

The center hole relief method is suitable for single wall part having center hole at the bottom. After axial upsetting, the formed part does not need more machining and can keep complete streamline. However, the effect of this method on reducing force is less than that of relief cavity. This is because the material has to flow to the center of the bottom, nevertheless, in the relief cavity method, the unnecessary material flow to relief cavity directly; the flow distance of the former method is longer than that of the later method. But the part manufactured by relief cavity method has to be machined to clear the unnecessary material away, which will break the streamline. In brief, it is necessary to take into account part structure and performance requirements when choose relief method.

2.3. Application

In this section, an application of axial upsetting is presented. The application object is flywheel plate used in self-changing gearbox. The dimensional drawing and three dimension model of part are shown in Fig. 7 and Fig. 8, respectively. It can be seen from figures that the flywheel plate is a cup type part with large diameter, 273.3 mm, while the wall thickness is 11 mm, and the bottom thickness is 10mm. The corner between wall and bottom is very small, inner round radius is 5 mm and outer round radius 2.5 mm. Moreover, there is one center hole and three auxiliary holes at the bottom of cup. The part material is 45 high quality carbon steel, which will has good mechanical properties after quenching and tempering.

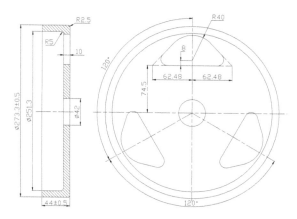

Figure 7. Dimensional drawing of flywheel plate.

Figure 8. Three dimensional model of flywheel plate.

Due to the part has different thickness in the wall and the bottom, and a small corner radius, traditional drawing process can not obtain desired part. If employing machining, the material usage is very low about 35%. Moreover, machining will cut streamline, which may lower mechanical properties. If the bottom and the wall are formed, respectively, and then combined to a complete flywheel plate by welding, the weld seam has a harmful effect on mechanical properties, which may not satisfy the performance requirement.

The thickening ratio of flywheel plate is 1.1. According to thickening ratio criterion, the wall thickness can be upset to designed value in one pass. So, the flywheel plate can be manufactured by sheet stamping-forging technology. Taking 10 mm circular plate as blank, firstly is drawn to a cup with uniform thickness, then the wall of cup is thickened to 11 mm by axial upsetting.

In order to decrease deformation stress, the warm forming was used in drawing and upsetting processes. Generally, the warm forming is a technology carried out at temperature between room temperature and recrystallization temperature, during which the deformation stress of material is significantly lower than that at room temperature.

The stamping-forging processes are as follows: blanking, heating, drawing, finishing inner corner and axial upsetting, piercing.

1. Blanking

    Due to the thickness of bottom is 10 mm, the circular blank with 10 mm thickness is chosen. The diameter of blank is 337.8 mm calculated by constant volume principle. To obtain high quality circular blank, fillet edge dies with small clearance between punch and die was used for blanking. The dimensional accuracy reached grade IT9-IT11, and the surface roughness was Ra3.2-0.8 μm.

2. Heating

    The electric resistance furnace full of protective atmosphere was used for heating the blank in order to reduce oxidization. The blank was heated to 800-850 °C and hold for 30 minutes. Meanwhile, the dies were heated to about 200 °C.

3. Drawing

    A 1000 kN mechanical press was employed to carry out the drawing process. The key process parameters, such as radius of punch and die, were determined by empirical principle.

    As blank thickness larger than 6 mm, the radius of punch can not less than 1.5-2 times of the thickness. In this application, the radius of punch must larger than 15-20 mm. So, the radius of punch was chosen as 15 mm to reduce the amount of finishing.

    The radius of die was determined by the following empirical equation:

    $$r_d = (2 \sim 4)t = (2 \sim 4) \times 10 = 20 \sim 40 mm$$

    where t is the thickness of blank (mm). In this application, the radius was 20 mm.

4. Finishing inner corner and axial upsetting

    An 8000kN mechanical press was used to finish inner corner and upset the wall. After the inner corner was ironed by a punch with 5 mm radius, the side wall was upset to desired dimension while the bottom of cup was clamped. Because the relief cavity reduces upsetting force significantly, the relief cavity method was adopted in this application.

5. Piercing

    After axial upsetting, four holes at bottom were formed by piercing.

    The flywheel plate manufactured by stamping-forging technology is shown in Fig. 9. It can be seen that there is no defect in forging surface, and the thickness of outer wall reaches the specified dimension.

Figure 9. Flywheel plate of car with outer wall thickening.

3. Thickening of inner wall of cup parts with axial upsetting

3.1. Design of thickening process for inner wall of cup parts

The forming process and methods of typically double-layer cup with inner wall thickened by axial upsetting are introduced in this section.

As shown in Fig. 10a, the traditional metal forming method for manufacture double-cup-shape part with thickened inner wall is that several partitions divided from the part is formed separately and assembled by the welding to a whole part. The part made by this method would decrease the mechanical performance of the part and production efficiency. In stamping-forging processing (see Fig. 10b), an initial blank is formed to a cup firstly by forward drawing. Then the bottom of the formed cup is drawn backward to form a double-layer cup by powerful drawing. While backward drawing, the material of outside wall is pushed to the inner wall making the inner wall as thick as possible. Subsequently, the center hole is formed by piercing and the inner wall is straightened by flanging. Finally the inner wall of the cup is thickened by upsetting.

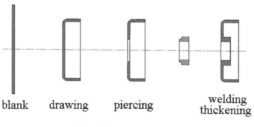

blank   drawing   piercing   welding
                                            thickening

a) Traditional processing

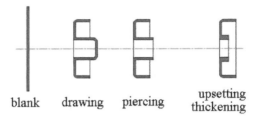

blank   drawing   piercing   upsetting
                                                   thickening

b) Stamping-forging processing

Figure 10. Scheme of forming process

3.2. Material store-up for inner wall using reverse drawing

In the forming of double-cup-shape part, the reverse cup-shape partition is formed by backward drawing to a certain height, which is prepared for subsequent thickening process. However, because it's difficult for the material of the outside wall flows to the bottom through the fillet, the backward drawing causes severe decrease in the thickness of inner wall. In order to improve stiffness of the inner wall and make the upsetting carried out latter more smoothly, the thickness of inner wall after backward drawing should be as thick as possible. Namely, it's necessary to store material during backward drawing.

As the required part could not be obtained by conventional backward drawing in one procedure, the process of powerful drawing in which a downward thrust is exerted on the outside wall to make metal flow to the inner wall is presented to get the thickened inner wall, as shown in Fig. 11. A pressure ring and a punch moves down and up, respectively, while the blank is clamped by blank holder. The die which is given an axial upside back pressure moves with pressure ring.

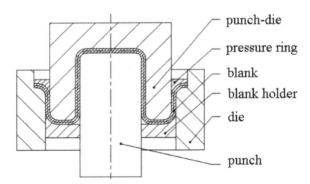

punch-die

pressure ring

blank

blank holder

die

punch

Figure 11. Diagram of thickening by powerful backward drawing

3.3. Application

3.3.1. Simulation of stamping-forging processing for the center hole edge of clutch hub

Stamping-forging processing for the center hole edge of car clutch hub was presented in this section. The schematic of clutch hub is shown in Fig. 12. As we can see in view B, the center hole edge is 3.5 mm in height and 1 mm in thickness, while in other position of part the thickness is 2.5 mm. There is a bevel of 45° to the center hole and a right angle shape around the hole. According to the features of the component, SFP technology should be employed. So, drawing, piercing, and spinning processes were used to obtain its structure of hollow shape, and then flanging and upsetting were used to obtain the required part.

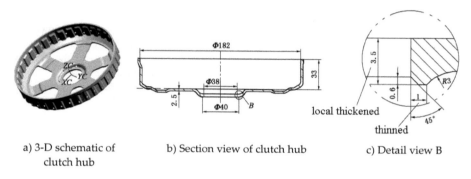

a) 3-D schematic of   b) Section view of clutch hub   c) Detail view B
   clutch hub

Figure 12. Schematic of clutch hub

As shown in Fig. 13, the shape of upper-punch and die cavity is in accord with the component, while the upsetting-punch (shown in Fig. 13b) is with ladder shape. Two steps were applied to implement the procedure: in the first step, the flanging-punch moved down to complete flanging process while other dies stayed where they were, and then it moved up; in the second step, the center hole edge was thickened and the right angle of it was formed with the upsetting-punch moving up. Since the thickness of the inner wall is not thinned by flanging, machining was needed to obtain the diameter 38 mm as well as the angle 45°.

The FEM software MSC.Marc can be used to simulate the forming process by taking half of the part because of the axisymmetric shape. Four-node quadrangles were available, and the mesh adaptive function was activated considering the large deformation in the process of forming. The initial thickness of the blank was 2.5±0.1 mm, and the stress-strain curve was obtained by tensile test. As we described before, the flanging-punch moved down and returned for the first 400 steps to finish flanging, and then in the last 200 steps of simulation, the upsetting-extruding punch moved up to complete upsetting process. The friction coefficient was set as 0.1.

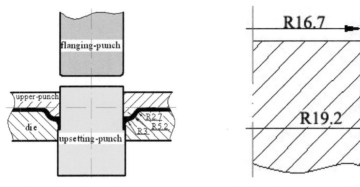

a) Scheme of punch and die of upsetting   b) Size of upsetting-extruding punch

Figure 13. Scheme of punch and die when upsetting

The result of finite element simulation is shown in Fig. 14. It can be observed that there is no folding at the bottom and the right-angle shape is formed perfectly. The load for the upsetting-extruding punch predicted by FEM is given in Fig. 15, of which the maximum is about 526.7 kN.

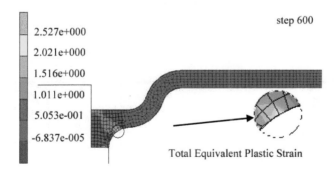

Figure 14. Result of upsetting-extruding

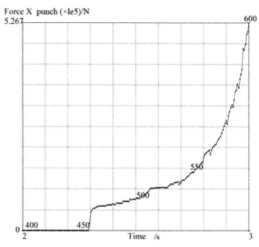

Figure 15. Forming load curve (from step 400 to step 600)

3.3.2. Experiment on stamping-forging process of double-cup-shape part with thickened inner wall

The typical double-cup-shape part with thickened inner wall is shown in Fig.16. It is a rotary part, in which the thickness of the inner wall is larger than that of other region. In order to improve the mechanical properties of the part, material utilization and production efficiency, the stamping-forging hybrid forming process mentioned before was used. As described earlier, the process was divided into three stages: forward drawing, then backward drawing, piercing and flanging, finally upsetting to get the inner wall thickened (shown in Fig. 17).

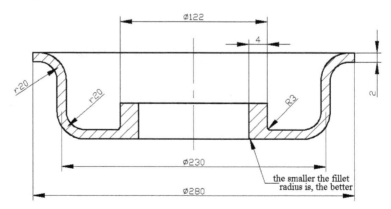

Figure 16. Scheme of double-cup-shape part

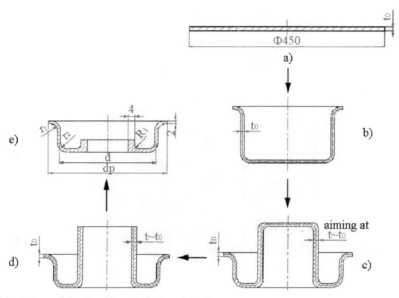

a) Blanking b) Forward drawing c) Powerful backward drawing
d) Piercing and flanging e) Thickening by upsetting

Figure 17. Scheme of stamping-forging processing of double-cup-shape part

The mechanical properties of the material are shown in Table 1. The experiment was conducted at a dual-action deep drawing hydraulic press (see Fig. 18), and the sheet metal material was 08AL steel with initial thickness of 2 mm. The nominal pressure of inner slider is 3000 kN and the outer is 2000 kN. The velocity and maximum effective stroke of the inner and outer slider are 10 mm/s and 500 mm, respectively. The nominal pressure of ejector of the hydraulic machine is 1000 kN, while the ejection stroke is 160 mm, and the velocity is 30 mm/s.

Parameter	Density/kg/mm^3	Young's modulus /N/mm^2	Poisson ratio	Yield stress /N/mm^2
Value	7.8×10^{-6}	2.07×10^5	0.28	1.713×10^2

Table 1. Mechanical properties of the material

The partial view of backward drawing and upsetting die is shown in Fig. 19. Δ is clearance which influences the flow of material in powerful backward drawing between punch-die and the die. If the gap is too much, wrinkling and folding defect may happen more easily. Instead, the resistance force will be increased, which leads to thinning or even rupture of the inner wall in backward drawing process. According to the thickness of the sheet metal is 2 mm, experiments on the gap were conducted more than once until the most satisfied clearance 3.5 mm was obtained. The formed part is shown in Fig. 20.

Figure 18. Dual-action deep drawing hydraulic press

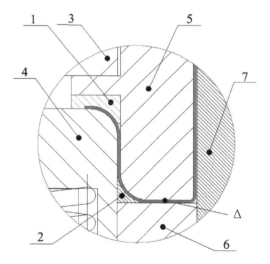

1- Upper pressure ring 2- Lower pressure ring 3- Blank holder
4- Floated die 5- Punch and die 6- Fixed die 7- Mandrel

Figure 19. Partial view of upsetting die

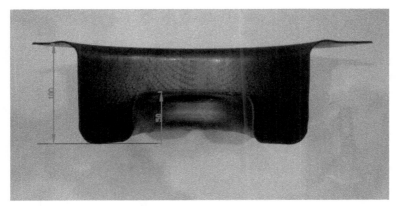

Figure 20. Section plan of double-cup-shape part by backward drawing

After powerful backward drawing, material of outer wall flowed into inner wall. The height of outer wall was decreased from 128 mm to 100 mm, while the height of inner wall was increased to 50 mm. Fig. 21 shows the measurement path of wall thickness along radial direction, and Fig. 22 shows the distribution of wall thickness. It can be seen that the wall thickness is not homogeneous along the radial direction, showing characteristics of shock fluctuation obviously. In the flange region from point 1 to point 10, the wall thickness is larger than that of initial blank due to composite effect of hoop pressure stress and tension stress in the radial direction during forward drawing. In the outer wall region from point 10 to point 23, the wall thickness is less than that of initial blank and it goes down from top to bottom. The diameter of the region point 23 to point 26 decreases with round corner of bottom, because this region is mainly suffered from hoop pressure stress resulting in thickening in the thickness. In the bottom region from point 26 to point 35, the wall thickness first decreases and then increases, and the thinnest place is at the center of bottom plane. In the inner wall region from point 35 to point 52, the wall thickness is severely thinned because of large tension stress in the radial direction; the closer near the central

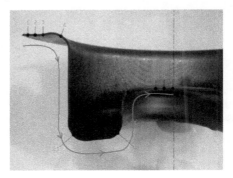

Figure 21. Measurement path of wall thickness of part

region, the thinner the wall is; and the thinnest point is the round corner of backward drawing punch. The region from point 52 to point 60 is also mainly suffered from tension stress, thus the entity wall is thinned. But close to the center, away from round corner of backward drawing punch, the blank is subjected to less deformation, so the wall thickness increases slightly compared with that of round corner.

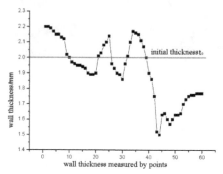

Figure 22. Distribution of wall thickness of part

According to the analysis above, thickness of inner wall is still thinned after powerful backward drawing due to various factors, such as stress in different region and friction between tools and material. However, if we do not use powerful backward drawing, the thickness of inner wall will be thinned more severely. Although the thinning of inner wall during backward drawing is not beneficial to the upsetting of this region, the double-cup-shape part with the inner wall of 4 mm was made successfully with optimized processing parameters. To avoid folding defect caused by bending of the blank, an upsetting with small gap and rigid support was used. The thickened part compared with non-thickened part is given in Fig. 23.

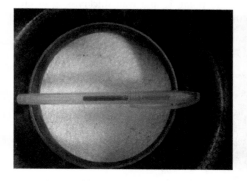

a) Part with inner wall non-thickened   b) Part with inner wall thickened by upsetting

Figure 23. Comparison of double-cup-shape part before thickening and after thickening

The forming process of the double-cup part is shown in Fig. 24. It is indicated that the double-cup part can be successfully formed by the mentioned stamping-forging hybrid processing.

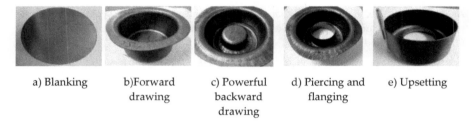

| a) Blanking | b)Forward drawing | c) Powerful backward drawing | d) Piercing and flanging | e) Upsetting |

Figure 24. Forming process of the part

4. Thickening of flange of disc-like parts with spinning

The forming process and methods of typically disc-like part with thickened rim thickened by spinning are introduced in this section.

The following calculations in section 4.1. and 4.2. are based on the assumption of plane deformation which the meridian plane of the work-piece keeps planar during forming process. That is, the deformation can be treated as a process of axisymmetric radial compression.

4.1. Design of multi-step process

Thickening ratio is also a key factor during the process design of thickening spinning. It is defined as $\lambda_n = t_n / t_{n-1}$, where t_n is the thickness of the rim after the n time thickening step, t_{n-1} is the thickness of the rim before the n time thickening step. The number of forming step required for rim thickening depends on the total thickening ratio $\lambda = t_N / t_0$, where t_N is the target thickness and t_0 is the initial thickness. Generally, the recommend value in a single thickening for low carbon steel is $\lambda_n \leq 1.4$. If $\lambda_n > 1.4$, the thickening could not be obtained in one forming step, a multi-step thickening process will be needed.

In a multi-step forming, t_n is decisive to the roller design and success of the process. Assuming the average strain in each forming step is equivalent, there is $\ln(t_1 / t_0) = ... = \ln(t_{N-1} / t_N)$. So, the t_n equals to

$$t_n = t_N^{n/N} t_0^{N-n} \qquad (1)$$

where, N is the total number of forming steps, n is the number of forming step, $1 \leq n \leq N$.

4.2. Radial feeding force calculation

The feeding force is critical to the choice and design of the capacity of the spinning equipment. An analytical model for calculating the forming forces is very useful, especially when a quick prediction of forming force is required.

As shown in Fig. 25, the final filled zone is the corner enclosed by roller, spindles and workpiece. A small sector body with thickness of one unit is analyzed to calculate the feeding force by slab method.

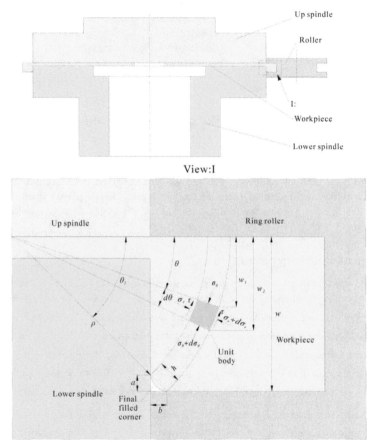

Figure 25. Sketch of principal stress on sector unit body

According to Fig. 25, the force equilibrium equation of the small sector body in θ direction is

$$\tau(2\rho + h)d\theta \times 1 + (\sigma_\theta + d\sigma_\theta)h \times 1 - \sigma_\theta h \times 1 = 0 \qquad (2)$$

Substituting boundary condition $\theta = \theta_1$, $\sigma_\theta = 0$, plasticity condition $\sigma_\theta - \sigma_r = \sigma_s = 2\tau$, and integrating σ_r along the cylinder surface, the mean feeding force f on the body of one unit can be expressed as:

$$f = \int_0^{\theta_1} \sigma_r \rho \cos\theta d\theta = \sigma_s \rho \left[\frac{m}{2} + \frac{m\theta_1}{2}\sin\theta_1 - \sin\theta_1 - \frac{m}{2}\cos\theta_1) \right] \tag{3}$$

where , $m = -(2\rho + h)/h$, $\theta_1 = \tan^{-1}(a/b)$, $h = \sqrt{a^2 + b^2}$, $\rho = (w-a)h/a$. h is the width of the sector body in radial direction, σ_θ and σ_r are stresses in tangential direction and radial direction, respectively. σ_s is the tensile strength, τ is the shear yield stress.

Fig. 26 shows the compressed zone of the workpiece, the total feeding force F can be expressed as

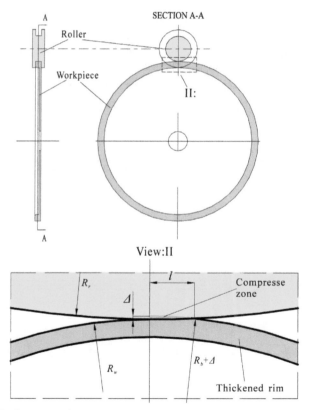

Figure 26. Sketch of compressed zone

$$F = \int_0^l p \, dt = \sigma_s \rho \left[\frac{m}{2} + \frac{m\theta_1}{2}\sin\theta_1 - \sin\theta_1 - \frac{m}{2}\cos\theta_1 \right] l \qquad (4)$$

where, R_r and R_w are the radius of roller groove and final workpiece, respectively. Δ is the feeding distance of the roller in one circle of the work-piece. The radii of the workpiece before the last circle is $R_w + \Delta$, according to Heron's Formula, the length of contacted zone is

$$l = \sqrt{(2r_r + 2r_w + \Delta)\Delta(2r_r - \Delta)(2r_w + \Delta)} / [2(r_r + r_w)]$$

The key to using the equation 4.4 is to obtain the values of a and b. In fact, it could be supposed that a equals to b, and set the value to be the allowable radii r_c of the required parts. Then $\theta_1 = \pi / 4$, $a = b = r_c$.

4.3. Application

In this section, an application example of SFP to manufacture a disc-like part of car with thickened rim will be introduced.

Fig. 27 shows a typical part manufactured by rim thickening. The rim thickness is 3.33 times to that of the center portion. The material is 1045 steel, whose Young's modulus is 210 GPa. The relationship of true stress to true strain at room temperature is $\sigma = C_\varepsilon^{-n}$ with C=1019.7 MPa and n=0.11, respectively. Firstly, a pre-formed part shown in Fig. 28 was made by stamping.

Figure 27. Sketch of part manufactured by rim thickening

Figure 28. Pre-formed parts by stamping

Because of $\sqrt[3]{\lambda} = \sqrt[3]{t_N / t_0} = 1.494 > 1.4$ and $\sqrt[4]{\lambda} = \sqrt[4]{t_N / t_0} = 1.351 < 1.4$, according to section 4.1, a four-step thickening process is required. The diameter of the spindle was 240 mm, which had the same value with the inner diameter of the thickened rim. The rollers' shape and parameters are shown in Fig. 29, two angular parameters c_1 and c_2 between the groove walls and middle plane, and a fillet with r_1 are designed to avoid scratch of the work-piece.

The middle span b_1 of the groove was calculated by equation (4.1). The parameter values of the tooling are given in Table 2.

Step No.	a	b_1	c_1	c_2	r_1	r_2	d
1	1.5	4.10	4	5	0.5	1.5	17.4
2	1.5	5.56	4	5	0.5	2.0	14.4
3	1.5	7.57	3	4	0.5	2.6	12.1
4	1.5	10.2	2	3	0.5	0	10

Table 2. Values of die dimension

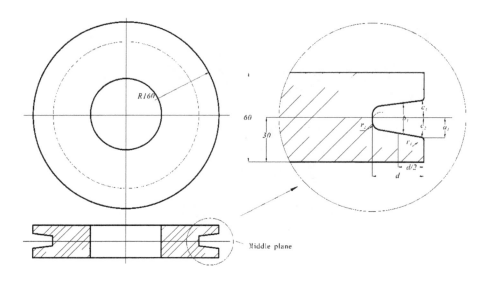

Figure 29. Shape of ring roller

The spinning machine with clamping capacity of 1000 kN was employed, as shown in Fig. 30. The tooling action is controlled by a PLC unit. The groove of the roller was heat-treated to hardness of HRC 58-62, and polished to surface roughness of 0.4 μm. The clamping force was set to be 500 kN during the rotary forming process. Graphite emulsion was used for lubricant and cooling. Feed speed was 0.05 mm per circle.

Fig. 31 shows the parts manufactured by multi-step spinning. According to section 4.2 and the shape of the final part, we can get $a = b = r_c = 1mm$, and the final feeding force is 86 kN calculated by equation 4.4 and 90 kN measured in experiment. The value of experiment is 4.6% higher than that calculated by equation 4.4.

Figure 30. Multi-step spinning machine

Figure 31. Parts with thickened rim made by multi-step spinning

Author details

Xin-Yun Wang, Jun-song Jin, Lei Deng and Qiu Zheng
State Key Laboratory of Materials Processing and Die & Mould Technology, Huazhong University of Science and Technology, Wuhan, China

5. References

[1] P. Jiang. Combined technology of cold stamping and forging for sheet metal and its application. Automobile technology and material. 2000,(9):8-11.

[2] P. Jiang, X. He, Y. Wu, H. Xie. New compound plastic forming technologies and their applications. Forging and stamping.2000,(1):38-41.

[3] C.J. Tan, K. Mori, Y. Abe. Forming of tailor blanks having local thickening for control of wall thickness of stamped products. Journal of materials processing technology. 2008,202:443–449.

[4] K. Mori, S.Nishijima, C.J.Tan. Two-stage cold stamping of magnesium alloy cups having small corner radius. International Journal of Machine Tools & Manufacture. 2009,49 :767-772.

[5] K. Mori, Y. Abe, K. Osakada, S. Hiramatsu.Plate forging of tailored blanks having local thickening for deep drawing of square cups.Journal of Materials Processing Technology. 2011,211:1569-1574.

[6] X.Y. Wang, M.L. Guo, J.C. Luo, K. Ouyang, J.C. Xia. Stamping-forging hybrid forming of double layer cup with different wall thickness. Materials Research Innovations, 2011,15(S1):435-438.

[7] X.Y. Wang, K. Ouyang, J.C. Xia, G.A. Hu. FEM analysis of drawing-thickening technology in stamping-forging hybrid process. Forging & stamping technology. 2009,34(4):73-78.

[8] X.Y. Wang, W.T. Luo, J.C. Xia, G.A. Hu. Investigation of warm stamping forging process for car flywheel panel. Forging & stamping technology. 2009,34(5):33-36.

[9] J.C. Luo, X.Y. Wang, M.L. Guo, J.C. Xia. Precision research in sheet metal flanging and upset extruding. Materials Research Innovations, 2011,15(S1):439-442.

Chapter 3

Developments in Sheet Hydroforming for Complex Industrial Parts

M. Bakhshi-Jooybari, A. Gorji and M. Elyasi

Additional information is available at the end of the chapter

1. Introduction

Sheet-metal forming processes are technologically among the most important metalworking processes. Products made by these processes include a large variety of shapes and sizes. Typical examples are automobile bodies, aircraft panels, appliance bodies, kitchen utensils and beverage cans [1].

Among the various sheet-metal forming processes, hydroforming is one of the non-traditional ones. This process is also called hydromechanical forming, hydraulic forming or hydropunch forming. In hydroforming process, liquid is used as the medium of energy transfer to form the workpiece. The part is formed on a female die, with the liquid under pressure acting in place of a conventional solid punch [1].

Hydroforming is applied more and more in the modern manufacturing industry. Comparing with the solid punch stretching process, hydroforming process results in a better strain state in the workpiece, so that a deeper draw can be achieved. The friction between tools and blank is greatly reduced. The advantages of hydroforming include low tooling cost, flexibility and ease of operation, low tool wear, no damage to the surface of the sheet, and capability to form complex shapes [2].

Types of hydroforming process

Hydroforming process is divided into two main groups; sheet hydroforming and tube hydroforming. These are briefly stated in the following sections.

Tube hydroforming process

Tube HydroForming (THF) is a process of forming hollow parts with different cross sections by applying simultaneously an internal hydraulic pressure and axial compressive loads to force a tubular blank to conform to the shape of a given die. Geometry of die and workpiece, initial tube dimension, tube anisotropy, and internal pressure are of the important parameters in this process [3].

With the advancements in computer control and high-pressure hydraulic systems, this process has become a viable method for mass production, especially with the use of internal pressure of up to 6000 bars. Tube hydroforming offers several advantages as compared to conventional manufacturing processes. These advantages include; a) part consolidation, b) weight reduction through more efficient section design, c) improved structural strength and stiffness, d) lower tooling cost due to fewer parts, e) fewer secondary operations (no welding of sections required and holes may be pierced during hydroforming), and f) tight dimensional tolerances. Despite several benefits over stamping process, THF technology is still not fully implemented in the automotive industry due to its time-consuming part and process development [4].

In THF, compressive stresses occur in regions where the tube material is axially fed, and tensile stresses occur in expansion regions. The main failure modes are buckling, wrinkling (excessively high compressive stress) and bursting (excessively high tensile stress). It is clear that only an appropriate relationship between internal pressure curve versus time, and axial feed curve versus time, so called Loading Paths (LP), guarantees a successful THF process without any of the failures [5].

Hydroformed tubular parts vary over a wide range of shapes. This variety goes from a simple bulged tube to an engine cradle with multiple part features such as bends, protrusions, and complex cross sections. It is necessary to classify the THF parts into different categories with respect to common characteristics that they have in order to handle the design process more efficiently [6]. Figure 1 shows some types of parts which are produced in this process.

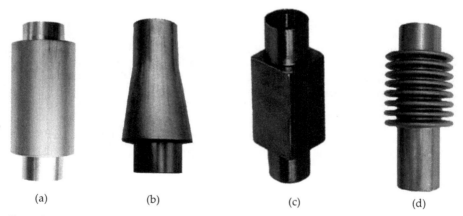

| (a) | (b) | (c) | (d) |

Figure 1. Tube hydroformed parts: (a) cylindrical stepped tube, (b) conical stepped tube, (c) rectangular stepped tube, (d) bellows [3-8]

Sheet hydroforming process

Sheet hydroforming process is an alternative to drawing process where either punch or die is replaced by hydraulic medium, which generates the pressure and forms the part. Sheet hydroforming is classified into two types Sheet HydroForming with Punch (SHF-P) and Sheet HydroForming with Die (SHF-D). In SHF-P (Figure 2(a)), the hydraulic fluid is replaced with the die, while in SHF-D (Figure 2(b)), the hydraulic fluid is replaced with the punch. Absence of either punch or die in SHF process reduces the tooling cost [9].

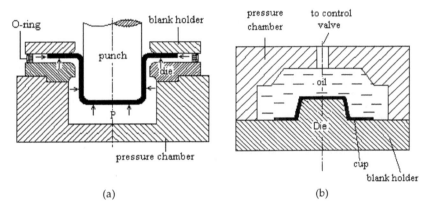

(a) (b)

Figure 2. Schematic of the sheet hydroforming process: (a) SHF-P, (b) SHF-D [9]

Types of sheet hydroforming process

Some new sheet hydroforming processes have been introduced during the last few years. Thiruvarudchelvan and Lewis [10], Anwar Kandil [11], and Zhang and Danckert [12] studied the standard hydroforming process. Figure 3 illustrates the tool set-up for this process. An essential part of this set-up is a rubber diaphragm that seals the liquid in the pressure chamber. During the process, the flange of the part is kept pushed against the blank holder by the fluid pressure transmitted through the diaphragm. There are many advantages for the standard hydroforming process, such as better surface quality and forming of complex shapes. Meanwhile, it has some disadvantages, such as the requirement of heavy presses. In addition, it is easy to destroy the rubber diaphragm, since the diaphragm is nearly under a similar deformation as the workpiece [10]. The fluid pressure is very important in this process, because wrinkles will appear if the pressure is not sufficiently high. If the pressure is too high, the blank may be damaged by rupture [12].

The hydromechanical deep drawing process has been developed by researchers on the basis of the standard hydroforming technology [10, 12]. Figure 4(a) shows the schematic illustration of this process. The fluid pressure in this process can be produced by the downward movement of the punch, or be supplied by a hydraulic system, since no rubber diaphragm is used. The tool device in this process is similar to that in a conventional deep drawing. An O-ring is used to prevent the flow out of the fluid on the flange. By using this process, more local deformation, increased drawing ratio and forming of complicated parts are realized [10]. But, in this method the blank holder force is not sufficient to prevent complex shapes from wrinkling. In addition, the fluid pressure is very high that causes high clamping force, and thus, heavy presses are required.

The hydromechanical deep drawing process has been developed as the radial hydromechanical deep drawing (hydro-rim) process (Figure 4(b)) [13]. In this process, when the punch goes down into the die cavity, the blank is forced into the die cavity filled with liquid. The liquid will be pressurized and will push the blank tightly on the punch surface. Also, the liquid pressure exists around the blank rim. This, in turn, can realize some forced radial feeding which is difficult in the current sheet hydroforming processes. In this method, since there is no clamping force (with rigid or semi-rigid part), wrinkling will occur easily in the forming of complex shapes.

In another investigation, Groche and Metz [14, 15] used an active-elastic blank holder system for high-pressure forming. Figure 5 illustrates the simplified schematic of this system. In this method, an elastic blank holder with a circular groove was used under the blank. The active-elastic blank holder system showed improvements with respect to the material flow in the flange area and reduced sheet thinning in critical corner regions of the workpiece. In this method the die set-up is complicated and the forming pressure is high.

Among the sheet hydroforming processes, hydrodynamic deep drawing assisted by radial pressure (HDDRP) has been used to form complex shapes and has a good drawing ratio [16]. This process is shown in Figure 6.

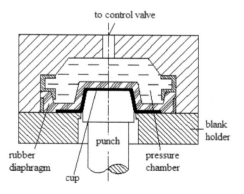

Figure 3. Tool set-up for standard hydroforming [9]

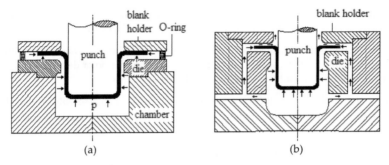

Figure 4. Tool set-up for: (a) hydromechanical, (b) hydro-rim, deep drawing process [9]

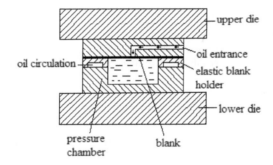

Figure 5. A simplified schematic of tool set-up for active-elastic blank holder system [9]

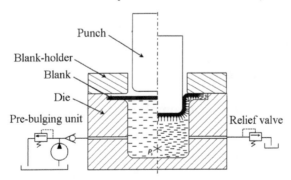

Figure 6. A simplified schematic of tool set-up for hydrodynamic deep drawing [16]

2. Applications and limitations of sheet hydroforming process [9-16]

Each of the sheet hydroforming mentioned above has its own limitations and applications. Sheet hydroforming process would be useful in reducing weight and cost simultaneously by improving structural integrity, strength and rigidity. In addition, this process satisfies these requirements with utilizing the common and available material efficiency.

Saving in tooling, material, design, production and assembly will altogether contribute reducing the overall cost of a sheet hydroforming part. Elimination or decrease of welds and welding operations is an additional of the overall cost.

A reduction in number of production steps and components in an assembly will be obtained with this process. This would reduce dimensional variations, and facilitate assembly operations.

Following is a list of potential advantages gained with the use of sheet hydroforming technology:

1. Reduction in weight
2. Increase in stiffness and rigidity
3. Economic material utilization
4. Complex shaped and various part types
5. Reduction in number of steps during manufacture and assembly (reduced welding and associated fixturing)
6. Reduction in overall cost per part or cost of assembly
7. Tight tolerances with good dimensional characteristics and less variation
8. Good surface finish

In contrast, longer process cycle and higher tool cost are limitations of this process.

2.1. Selection of a sheet hydroforming process for a complex industrial workpiece

The range of the applications of any sheet hydroforming process is limited. Not all the processes can be used for complex industrial parts. In this section, the criteria in selecting a specific process will be explained, such as the high drawing ratio, control of wrinkling, and ease of applying internal pressure.

3. Case study: Introducing two complex industrial workpieces

Case study I [9]

Figure 7 shows the photograph of the product that is used in lightning industry. The part is made of Al 1100 sheet with a thickness of 1mm and with strain hardening equation of $\sigma=610$ $\varepsilon^{0.24}$ [MPa]. Currently, it is produced in the collaborating industry by the conventional deep drawing process in several stages.

Three main features of the workpiece are:

1. It has a very complex shape with special internal profile that should be produced with high precision and surface finish.

2. There is a sharp edge on the workpiece that is specified in Figure 7(a). This is the most complex region of the workpiece to be formed.
3. The flange area of the workpiece that is shown in Figure 7(b), should be produced with good accuracy and with no wrinkle.

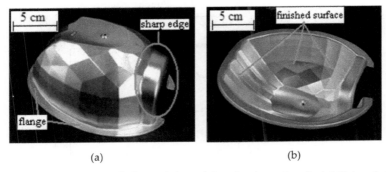

(a)   (b)

Figure 7. Case study 1; The part of lightning industry (after trimming and coating): (a) internal view, (b) external view

Case study II [17]

Drawing of conical parts is considerably more difficult than the deep drawing of cylindrical cups. Conical shape forming through conventional deep drawing is shown in Figure 8. As it is shown in the figure, due to small contact area of the punch tip with the blank at initial stages of deformation, high stresses are applied to this area of the sheet. This may cause bursting in the sheet. In addition, when forming conical cups through conventional deep drawing, wrinkling occurs on the sheet wall because the blank is free between the punch and the die [18, 19]. Thus, conical parts are normally formed by multi-stage deep drawing [18], spinning [20] or explosive forming [21].

Because of the large radial tensile stresses, small drawing ratios must be used in each stage when making conical components in multi-stage deep drawing. In addition, the ratio of the sheet thickness to the initial blank diameter influences the limiting drawing ratio to a greater extent than when drawing cylindrical parts. The limiting drawing ratio also depends on the cone angle and the ratio of the largest to the smallest cone diameter [18].

Experimental set-up

In all the hydroforming processes a universal testing machine (Figure 9(a)) and a hydraulic unit (Figure 9(b)) are used. In addition, the hydraulic unit was used as the pressure medium to form the workpiece. A control valve with a maximum pressure of 50MPa regulated the liquid flow to maintain the required pressure.

Tool set-up: Case study I

Figure 10 illustrates the schematic illustration of the new tool set-up proposed for this part. Figure 11 shows the photograph of the punch used for the part. The manufactured die-set is

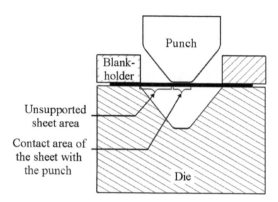

Figure 8. Schematic illustration of forming a conical part in conventional deep drawing

Figure 9. (a) Testing machine, (b) hydraulic unit

shown in Figure 12. It consists of a punch, a blank holder, a steel sheet ring, a die (oil container), a rubber diaphragm and an O-ring, which seals the liquid in the pressure chamber (die). As it is seen from the figure, the rubber diaphragm is only used in the region between the blank holder and die. Therefore, no diaphragm is used in the deformation region of the blank. In this region, the liquid is in direct contact with the blank. Thus, while the diaphragm is prevented from any deformation and tearing, lower pressure is required to form the workpiece, compared with the case in the standard hydroforming.

In contrast to the conventional deep drawing, the sheet metal is not in direct contact with the die in the proposed method. Also, the blank holding system in this method is a soft-tool one. Therefore, wrinkles can be controlled to high extent, compared with the case in the hydromechanical and hydro-rim processes.

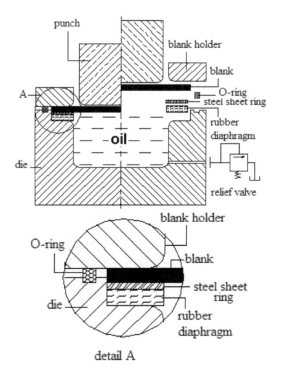

Figure 10. Schematic illustration of the tool set-up in the proposed method

Figure 11. Photograph of the manufactured punch for case study I

The operational sequence for the new hydroforming process is given below:

1. The die was clamped on the machine anvil. The rubber diaphragm was located on the die, the steel sheet was put on the rubber, and the O-ring was located in the groove

machined in the die. Then, the die cavity was filled with the pressure medium. The circular blank was then located on the steel sheet ring.

2. The blank was clamped between the die and blank holder by four screws. At this stage, an initial blank holder force was applied to the blank and a pre-hydroforming pressure was exerted on the lower surface of the rubber diaphragm and blank, lead to a small preloading on the blank.

3. The punch that was attached to the machine ram, moved down, pressurized the medium, and the deformation was started.

Figure 12. Photograph of the manufactured die set

Figure 13(a) illustrates the variations of internal pressure with punch stroke for the hydroforming of part in case study 1. As it is seen from the figure, the maximum forming pressure is about 5.5 MPa that is very low, compared with the results of reference [16], which formed simple parts with other hydroforming processes. As it can be seen from Figure 13(a), the internal pressure in the final stage of hydroforming oscillates. This is due to forming the special internal profile of the workpiece. Figure 13(b) illustrates the load-punch stroke curve of the workpiece. As it is seen from the figure, the maximum load is about 60kN which is not so high.

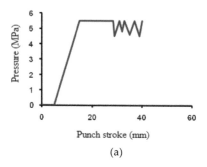

(a)

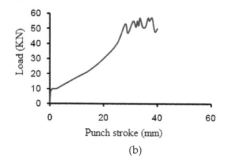

(b)

Figure 13. (a) Internal pressure-punch stroke curve, (b) load –displacement curve, correspond to case study I

Tool set-up: Case study II

In this study, pure copper and St14 sheets with different thicknesses and different initial diameters were used to form conical-cylindrical cups in one stage by HDDRP process. Mechanical and physical properties for the sheets are shown in Table 1. Based on the experimental observations, the copper sheet did no behave any anisotropy. To characterize the material properties and anisotropy for steel sheets, according to ASTM-A370 standard different specimens were cut at different orientations to the rolling directions (0°, 45°, and 90°). Tensile specimens were used to determine the stress–strain curves and the sheet anisotropy parameters, r-values. Plastic strain ratio (r-value) and yield stress ratio (R-value) for St14 sheet are illustrated in Table 2.

The schematic of the hydroforming die set used in this research is shown in Figure 14. Figure 15 shows the parametric geometry of the workpieces examined. Initially, three types of blanks with similar geometries but with variation in material or sheet thickness were selected. The specifications of these blanks and the formed parts are shown in Table 3 as parts A, B and C and with strain hardening equation of $\sigma = k\,\varepsilon^n$, where k and n are specified in the table.

A component with different geometry from those of the above mentioned parts was also selected to have a wider study. This is specified in Table 3 as part D.

Blank material	Young's modulus, E (GPa) [16]	Strain hardening exponent, n	Strength coefficient, K	Yield stress, σ (MPa) [16]	Poisson's ratio [16]	Density, ρ (kg/m³) [16]
Cu 99.9 %	117	0.44	530.98	123	0.32	8940
St 14	210	0.35	638.96	190	0.3	7850

Table 1. Mechanical and physical properties for pure copper and St14 sheets

Thickness (mm)	R0	R45	R90	R22	R33	R12
1	1.79	2.27	1.01	1.0402	1.24897	1.0789

Table 2. Plastic strain ratio (r-value) and yield stress ratio (R-value) for St14 sheet

A DMG (Denison Mayes Group) universal testing machine with 600kN capacity was used in the experiments.

Figure 16 shows the components of the die and the assembled die-set. Figure 17 shows the parametric dimensions of the die set used for forming the parts. The dimensions of the die set are given in Table 4.

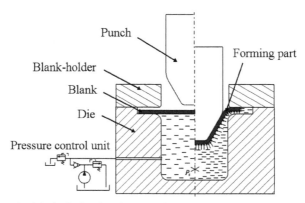

Figure 14. Schematic of the hydroforming die set used

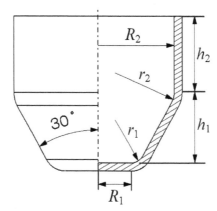

Figure 15. Parametric dimensions of formed parts

Type and dimensions of the part	Values related to parts	
	Parts (A, B, C)	Part (D)
Type of material	A, B=Pure Copper, C= St14	Pure copper
Height of conical portion, h_1	20	40
Height of cylindrical portion, h_2	18	16
Flat head radius, R_1	8	7.5
Cylindrical radius section, R_2	20.75	35.35
Nose radius, r_1	3.5	10.5
Conical-cylindrical radius, r_2	5.5	6.5
Initial blank thickness t_0	A=2, B=1, C=1	2

Table 3. Parameters of the part corresponds to Figure 15 (Dimensions in mm)

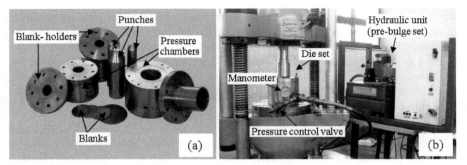

Figure 16. (a) components of the die, (b) assembled die set mounted on the test machine

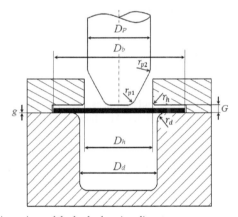

Figure 17. Parametric dimensions of the hydroforming die set

Parameter	Values related to parts	
	Parts A, B, C	Part D
Die inside diameter, D_d	A=46, B=C=44	75
diameter, D_h Blank holder inside	42	71
Blank diameter, D_b	78	120
Blank holder entrance radius, r_h	3	3
Die entrance radius, r_d	4	5
, G Gap between die and blank-holder	A= 2.2, B=C=1.2	2.2

Table 4. Parameters of the die corresponds to Figure 17 (Dimensions in mm)

In the HDDRP, before the punch goes down, a small pre-bulging can be created on the sheet to improve the drawing process [16]. A hydraulic unit was used to create the pre-bulging pressure. When the punch moves down, the blank is forced into the die cavity filled with oil or other liquids. A control valve was used for controlling the maximum pressure. The liquid in the die cavity was pressurized so it pushes the blank tightly onto the punch surface. After reaching a maximum pressure, the control valve was opened and the pressure remained constant during the forming process. The liquid in the die cavity leaks out dynamically from the interface between the blank-holder and the die. The interface between die and blank-holder was grinded metal contact and no o-ring was used in the die. At the same time, the liquid leaking out from this interface creates a pressure around the outside rim of the blank. Therefore, it is impossible to create high pre-bulging pressure in this die set. In this research, 2 MPa pre-bulging pressure was applied.

Figure 18 shows the typical pressure path used in this study. In this path, OA is the initial pre-bulging pressure (2 MPa) applied before the punch moves down. BC is the constant maximum pressure. The liquid outflows from control valve by applying this pressure. SAE10 hydraulic oil with a viscosity of 5.6 cSt was used as the pressure medium. Due to the strain-rate sensitive behavior of the viscous medium, the punch velocity has significant effect on the internal pressure generation. Thus, in the pressure path of Figure 18, AB is the linear pressure path and its slope depends on punch velocity and workpiece shape and thickness. In this research, a punch velocity of 200mm/min was applied. To measure the cup thickness, a mechanical thickness measurement set was used.

The typical pressure path in this paper was shown in Figure 18. According to the figure, for each certain maximum pressure, a pre-bulging pressure, OA, and a pressure path AB with different slopes are definable. The slope of AB changes with punch velocity, workpiece shape and sheet thickness. The punch velocity was fixed at 200mm/min. Thus, for each certain part with defined shape and thickness, one specific slope was obtained.

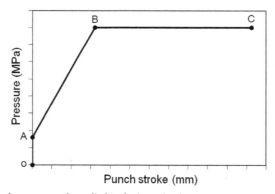

Figure 18. The typical pressure path applied in the investigation

FEM simulation

The commercial software, ABAQUS 6.7/Explicit, was used for the simulation. For pure copper sheet, the material behavior was assumed to be isotropic as the experimental results have verified this assumption. For St14 sheet, the anisotropy factors mentioned in the previous section were used in the simulation. 3D models were used for the simulation. The blank was modeled deformable with eight-node solid element (C3D8R). The number of elements along the thickness was 4. The die set was modeled using a rigid four- node shell element (R3D4).

The die and the blank holder were constrained fully and the punch could move only along the vertical direction, corresponding to the central axis of the punch. Pressure constrains were applied on the whole bottom surface and also on the rim of the blank. The gap between the die and the blank holder was fixed. The punch motion was prescribed with a constant velocity. Because of the consideration of pre-bulging, the loading of liquid pressure in the die cavity was used as a two-step linear profile. The friction coefficient on the blank and the punch interface was considered to be 0.14 in the simulations. The coefficient on the other surfaces was considered to be 0.04. Penalty contact interfaces were used between the sheet metal and the tooling elements. Table 1 shows the properties of pure copper and St14 sheets which have been used.

4. Results and discussion

Case study I

The schematic of the modified die-set for case study I is shown in Figure 19. The photograph of the used punch is shown in tool set-up section. To form this part, several pressure paths have been examined by FE simulation and the appropriate pressure path is shown in Figure 20. As it can be seen in the figure, the maximum forming pressure is about 5.5MPa which is very low, in comparison to the results of the other relevant references, which formed simple parts with other hydroforming processes.

Figure 21 shows the photograph of the workpiece formed in the new die-set. The initial blank is a round one with a diameter of 140 mm. As it can be seen in this figure, the workpiece is formed quite well to the final required height, only in one step. The sharp region of the workpiece is formed successfully. The internal surface of the product is formed with high precision and good surface finish and there is no, even one small, wrinkle on the flange area.

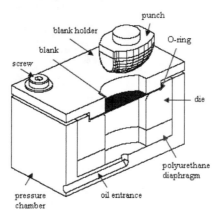

Figure 19. Schematic illustration of the proposed set-up for a complex part

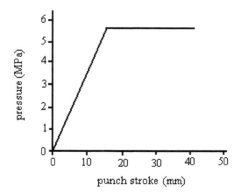

Figure 20. Pressure path used to form the lightning industry product

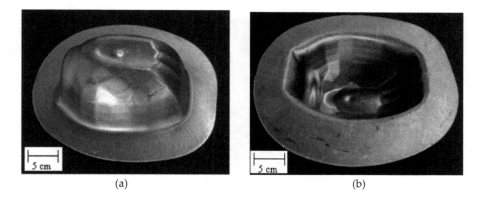

(a)   (b)

Figure 21. Photograph of the hydroformed part in the proposed die-set, (a) external view, (b) internal view

Case study II

Figure 22 shows the desired paths corresponding to maximum pressures for parts A, B, C and D. For parts B and C the punch velocity, workpiece shape and sheet thickness are the same which leads to the same slope.

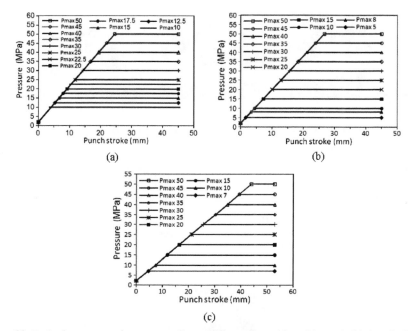

(a)

(b)

(c)

Figure 22. Desired pressure paths corresponding to different P_{max}, to form (a) part A, (b) parts B, C, (c) part D

The results obtained from experiment and simulation illustrated that for part A with maximum pressures of less than 12.5 MPa bursting occurs in the contact area of the workpiece with punch nose radius. Figure 23 shows a model of part A formed with maximum pressure of 10 MPa along with its simulation results. As it can be seen, bursting has occurred in the workpiece because the low level of forming pressure leads this process to act just like as the conventional deep drawing.

The results indicated that in parts B, C, and D bursting occurs in the same area at maximum pressures less than 7.5, 17.5 and 7.5 MPa, respectively. Figure 24 shows the picture of parts B, C and D in which bursting occurred.

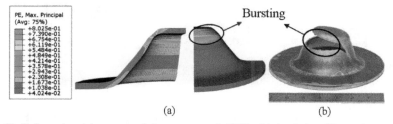

(a)   (b)

Figure 23. Deformed part A corresponds to pressure path 10MPa, (a) simulation, (b) experiment

(a)   (b)   (c)

Figure 24. Deformed workpieces correspond to different pressure paths, (a) part B, P_{max} = 5 MPa, (b) part C, P_{max} =15 MPa, (c) part D, P_{max} = 5 MPa

In part A with maximum pressure above 12.5 MPa the conical cup was formed. Figure 25 shows the picture of part A formed with maximum pressure of 12.5 MPa. As it is seen, at the end of the process the conical cup was formed, but necking occurred in the workpiece.

The results indicated that in parts B, C, and D necking occurred in the same area at maximum pressures of 7.5, 17.5 and 7.5 MPa, respectively. Figure 26 indicates the picture of the parts in which necking occurred.

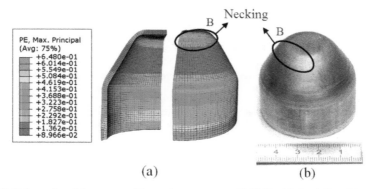

(a)   (b)

Figure 25. Deformed part A correspond to maximum pressure 12.5 MPa, (a) simulation, (b) experiment

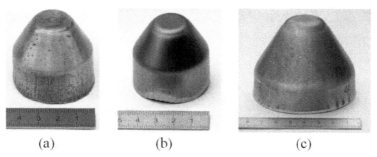

(a)   (b)   (c)

Figure 26. Deformed workpieces corresponds to pressure paths, (a) part B, P_{max} = 7.5 MPa, (b) part C, P_{max} = 17.5 MPa, (c) part D, P_{max} = 7.5 MPa

It was observed that increasing the maximum pressure leads to decrease necking. Figure 27 indicates the formed part A corresponding to 25 MPa maximum pressure. Applying this maximum pressure at final stage causes complete forming of the cup without necking defect and creates an accurate geometry.

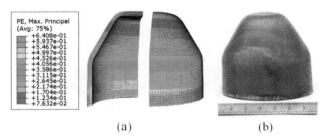

(a)   (b)

Figure 27. Deformed part A corresponds to pressure path 25 MPa, (a) simulation, (b) experiment

Parts B, C and D were formed without any necking occurrence with the maximum pressure paths of 17.5, 35 and 20 MPa, respectively, and are shown in Figure 28.

To have more careful study of thickness distribution, the deformed cups were divided into different regions as it is shown in Figure 29.

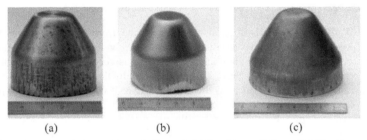

(a)   (b)   (c)

Figure 28. Deformed workpieces, (a) part B, P_{max} = 17.5 MPa, (b) part C, P_{max} = 35 MPa, (c) part D, P_{max} = 20 MPa

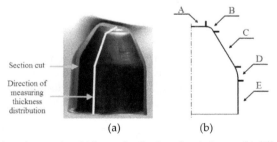

(a)   (b)

Figure 29. (a) Direction of measuring thickness distribution of conical parts, (b) different regions on the formed part

Figure 30 shows the thickness distribution curve for part A with maximum pressure of 25 MPa. As it is seen from the figure, there is a good correlation between the results of simulation and experiment.

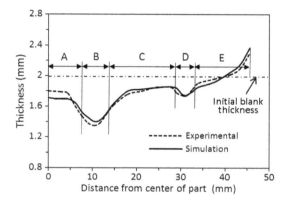

Figure 30. Thickness distribution curves for part A in maximum pressure of 25 MPa

In Figure 31 the thickness distribution curves obtained from experiment for different maximum pressures for four parts are displayed. As it can be found from this figure in the top of the conical cup, region A, the thickness reduction is very small. The most thickness reduction occurred in B and D regions. This thickness reduction is because of the bending occurrence in these regions. Region B is the critical zone as it was indicated in the previous results. In C and E regions the thickness increases and this thickness increase becomes greater toward the edge. It is obvious that the pressure increasing has a great effect on thickness reduction at different points of cup, especially in the critical region B.

For obtaining the best forming pressure path to produce a cup with better thickness distribution and quality, the maximum thickness reduction curve in B region was compared for different pressures. As stated previously, this is the most critical region of conical formed parts. Figure 32 shows the thickness distribution curves in B region corresponding to pressure paths with different maximum fluid pressures. As it can be found in Figure 32 (a), in part A the greatest thickness reduction is related to maximum pressure path 12.5 MPa. At this pressure, necking defect occurred in region B. It can be seen in the figure that by increasing the maximum pressure to 25 MPa, thickness reduction decreases with sharp slope. From the maximum pressure of 25 MPa and greater, the slope will not change considerably. Thus, maximum pressure more than 25MPa does not have any positive effect on the cup thickness in region B. Also, in Figure 32 (b), (c), (d) it can be found that the similar behavior happened for other parts but the greatest thickness reduction and minimum thickness reduction are different.

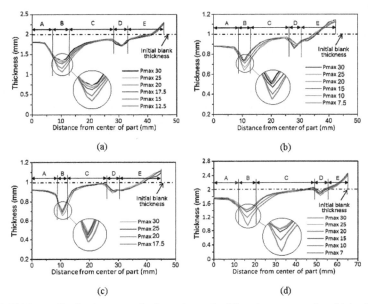

Figure 31. Thickness distribution curves for formed parts in different pressure paths, obtained from experiments, (a) part A, (b) part B, (c) part C, (d) part D

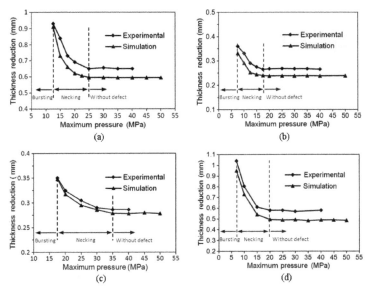

Figure 32. Thickness reduction curve versus maximum pressure for region B of conical parts, (a) part A, (b) part B, (c) part C, (d) part D

The drawing ratio of conical part is the ratio of initial blank diameter to the minimum diameter of the conical potion. The drawing ratio for parts A, B and C is 4.875 and for part D is 8. The relation between the drawing ratio and the lowest maximum forming pressure has been studied through simulation which is shown in Figure 33. In this figure it can be seen that with the sheet diameter decreasing, the conical part forming will be possible at lower pressures. When the sheet diameter or drawing ratio is increased, the forming pressure increases too, but with increasing the pressure, the sheet diameter increases to some extent. In the workpiece A, the sheet with the maximum of 87 mm in diameter can be formed through increasing the pressure but forming a blank with a diameter greater than 87 mm is not possible.

In Figure 33 the maximum drawing ratio for parts A, B, C and D are 5.4, 6.06, 5, and 9.67, respectively. In conical part forming through HDDRP, the sheet thickness, the conical angle, punch tip radius (B region) and sheet properties have great and considerable influence on the drawing ratio.

Punch force is related to the forming force and internal pressure in vertical direction. As it can be seen in Figure 34, the punch force increases when it moves down to reach a maximum force, and as the punch continues to move downward the punch force decreases.

By increasing the maximum pressure, the more punch force is needed. This is because the vertical direction force increases. Figure 35 illustrates the effect of maximum fluid pressure on punch force for parts A, B, C and D.

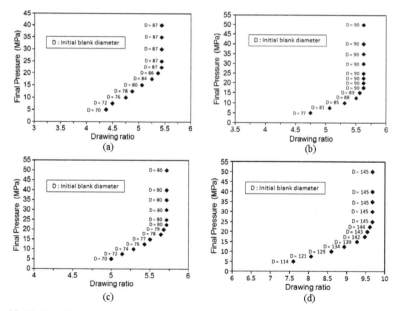

Figure 33. The drawing ratio corresponding to maximum pressure, P_{max}, (a) part A, (b) part B, (c) Part C, (d) part D

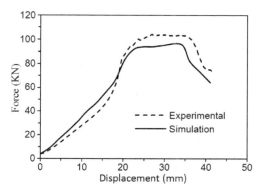

Figure 34. Force–punch stroke curve

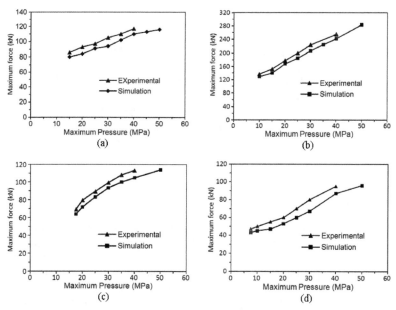

Figure 35. Maximum punch force versus maximum pressure corresponding to (a) Part A, (b) Part B, (c) Part C, (d) Part D

As it was stated, a pressure increase leads to a thickness reduction and on the other hand pressure increase makes the punch force more and there is a need to have bigger tonnage of press. So, obtaining the optimum pressure and punch force has a great importance in forming a workpiece with high quality.

To examine the effect of cone angle, different punches with 45^0, 60^0 and 75^0 angles were manufactured for part D geometry. Figure 36 shows %thinning in B region corresponding to different pressure paths for conical part with different cone angles. As it can be found, in conical part with 45⁰ angle, the greatest thickness reduction is related to maximum pressure 18MPa. At maximum pressure less than 18MPa bursting occurs in B region. It can be seen in the figure that by increasing the maximum pressure to 25MPa, thickness reduction decreases with sharp slope. From the maximum pressure of 25MPa and greater, the slope will not change considerably. In conical workpiece with 75⁰ angle, without applying any pressure, no bursting was observed. With increasing the pressure to almost 20MPa the thickness reduction decreases sharply. At maximum pressure of approximately 20MPa and beyond this, the slope will be horizontal. Figure 36 shows %thinning for the three different angles. It can be observed that with increasing the conical angle the thickness reduction will be decreased in B area. Moreover, as the conical angle increases, bursting occurs at lower pressure in such a way that beyond one specific angle, say 75⁰, the conical workpiece can be formed in the die chamber without applying any pressure.

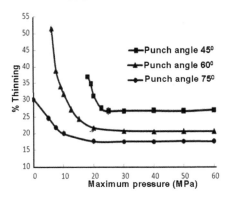

Figure 36. % thinning for different cone angles

Figure 37 shows the failure, thinning and safe forming regions for different conical angles. As shown in the figure, by increasing the punch conical angle, the bursting region becomes smaller. At angle of 75⁰ and higher the part did not fail even without applying the fluid pressure. In other words, by decreasing the conical angle, the pressure level should increase to prevent the failure. This analysis is valid for safe forming limits too. It means that by increasing the angle, the maximum pressure level decreases for forming an accurate part without any defect. At the angle of less than 35⁰, bursting occurs in the part at any fluid pressure.

Figure 38 shows the effect of punch friction coefficient on thickness distribution of the conical part. As it is shown in the figure, changes in the punch friction coefficient only affect region A and B, and it has no significant effect on other areas. It was observed that with increasing the punch tip radius the thickness reduction decreases in this region.

For more detailed review, the effect of punch friction coefficient in region A and B were studied for different punch friction coefficients.

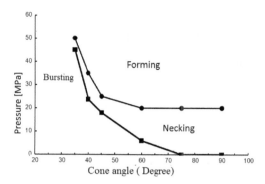

Figure 37. % thinning versus cone angle

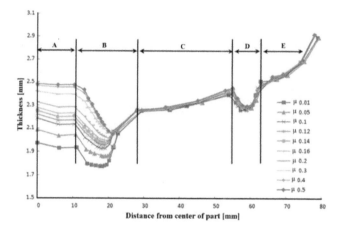

Figure 38. Thickness distribution curves versus punch friction coefficient, punch angle 60°

Figure 39 shows %thinning curves in A and B regions corresponding to different punch friction coefficients. From the figure it is observed that as the punch friction coefficient increases the thickness reduction decreases in the two regions. It can be seen in the figure that by increasing the punch friction coefficient to 0.3, thickness reduction decreases with sharp slope. From the punch friction coefficient greater than 0.3 the slope will not change considerably specially in B region.

Figure 40 illustrates the effect of the blank holder friction coefficient. As it shows, changes in the blank holder friction coefficient affect all regions. For more accurate study, the thickness reduction in B area was investigated and the results are shown in Figure 41. It is observed that by increasing the blank holder friction coefficient, the thickness reduction increases. At higher blank holder friction coefficients, necking occurs in B region. In this research, the blank holder friction coefficient of more than 0.3 results in bursting in region B.

Figure 42 shows the effect of sheet thickness on the thickness reduction in region B. As it is obvious, up to the maximum pressure of 20MPa the changes in the sheet thickness affect the thickness reduction. From the maximum pressure of 20MPa this effect is not considerable. Also, it can be observed that by decreasing the sheet thickness the possibility of thickness reduction reduces.

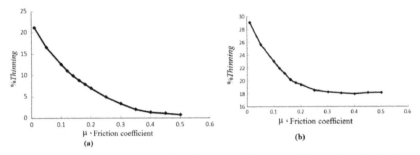

Figure 39. % thinning of the conical part for different punch friction coefficient, (a) region A, (b) region B.

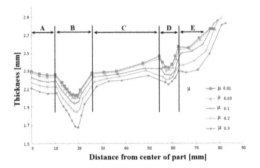

Figure 40. Thickness distribution curves for different blank holder friction coefficient, punch angle 60°

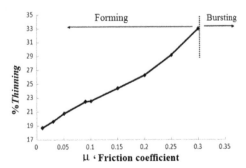

Figure 41. % thinning versus blank holder friction coefficient in region B

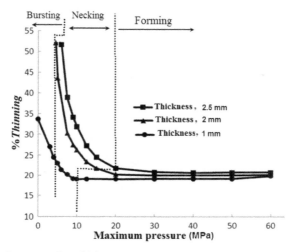

Figure 42. % thinning versus sheet thickness

Bending radius is a very effective factor on thickness distribution. Figure 43 shows the effect of punch tip radius on thickness distribution of the conical part. As it is shown in the Figure, changing the punch tip radius only affects region B and it has no significant effect on other areas.

Figure 44 shows %Thinning at different punch radiuses. As it can be observed, with increasing the punch tip radius the thickness reduction decreases in B region. With increasing the conical angle the thickness reduction will be decreased in region B. Moreover, as the conical angle increases, bursting occurs at lower punch radius in such a way that beyond one specific degree angle, say $75°$, the conical workpiece can be formed in the die chamber without applying any punch radius.

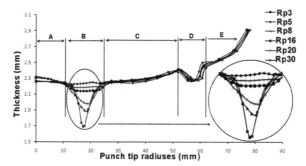

Figure 43. Thickness distribution curves versus punch tip radius, punch angle $60°$

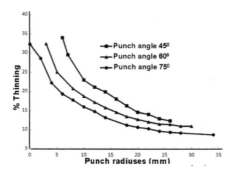

Figure 44. % thinning versus punch tip radius for region B

Figure 45 illustrates the effect of the radius of conical-cylindrical region on the thickness distribution. As it shows, changes in the radius only affects region D and it has no effect on other regions of the part. For more accurate study, the thickness reduction in region D was investigated and the results are shown in Figure 46. As it can be seen, by increasing the radius of region D and the conical angle the thickness reduction will be decreased.

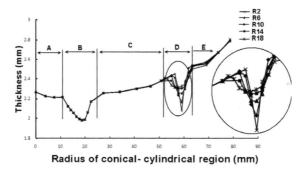

Figure 45. Thickness distribution curve versus radius of conical–cylindrical region

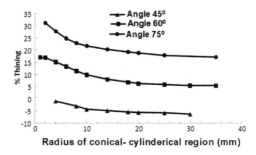

Figure 46. % thinning versus radius of conical-cylindrical region

5. Conclusions

In this chapter, a new sheet hydroforming is proposed and applied for forming of two industrial parts that currently are produced in industry by conventional deep drawing and stamping in several stages. With the new method, these two parts were produced in one stage and without any defects. In addition, it is shown that the forming pressure and load are very low compared with those of other hydroforming methods.

In addition, for case study II, the effects of tool parameters such as the radius of the punch tip, punch-cylindrical radius, friction between punch and sheet, friction coefficient between blank holder and sheet, sheet thickness and punch angle, on formability and thickness distribution of the conical parts were studied through using hydrodynamic deep drawing assisted by radial pressure. It was observed that with increasing the conical workpiece angle the thickness reduction will decrease in B area. Moreover, as the conical angle increases, bursting occurs at lower pressure in such a way that beyond one specific angle, the conical workpiece can be formed in the die chamber without applying any pressure. Also, with the cone angle increasing, the thickness distribution will be improved and the likelihood of bursting decreases.

Author details

M. Bakhshi-Jooybari, A. Gorji and M. Elyasi
Faculty of Mechanical Engineering, Babol University of Technology, Babol, Mazandaran, Iran

Acknowledgement

The authors would like to thank Dr. M. Hosseinzade for the provision of valuable information and product data.

6. References

[1] Singh H (2003) Fundamental of hydroforming, SME, 219 p.
[2] Koc M (2008) Hydroforming for advanced manufacturing, Woodhead publishing limited, 396 p.
[3] Elyasi M, Bakhshi-Jooybari M, Gorji A (2009) Mechanism of improvement of die corner filling in a new hydroforming die for stepped tubes. Materials and Design. 30: 3824-3830.
[4] Elyasi M, Bakhshi-Jooybari M, Gorji A, Hossinipour SJ, Norouzi S (2009) New die design for improvement of die corner filling in hydroforming of cylindrical stepped tubes. Proc. IMechE, Part B: J. Engineering Manufacture. 223: 821-827.
[5] Bakhshi-Jooybari M, Elyasi M, Gorji A (2009) Numerical and experimental investigation of the effect of the pressure path on forming metallic bellows. Proc. IMechE, Part B: J. Engineering Manufacture. 224: 95-101.

[6] Elyasi M, Bakhshi-Jooybari M, Gorji A, Hossinipour SJ, Norouzi S (2008) Numerical and experimental investigation on forming metallic bellows in closed and open die hydroforming. Steel Research International. 79: 148–154.

[7] Khanlari H, Elyasi M, Bakhshi-Jooybari M, Gorji A, Davoodi B, Mohammad Alinegad G (2010) Investigation of Pressure Path Effect on Thickness Distribution of Product in Hydroforming Process of SS316L Seamless Tubes. Steel research international. 81: 560-563.

[8] Elyasi M, Zoghipour P, Bakhshi-Jooybari M, Gorji A, Hosseinipour SJ, Nourouzi S (2010) A New Hydroforming Die Design for Improvement of Die Corner Filling of Conical Stepped Tubes. Steel research international. 81: 516-519.

[9] Hosseinzade M, Mostajeran H, Bakhshi-Jooybari M, Gorji A, Norouzi S, Hossinipour SJ (2009) Novel combined standard hydromechanical sheet hydroforming process, IMechE Journal of Engineering Manufacture, 224: 447-457.

[10] Thiruvarudchelvan S, Lewis W (1999) A note on hydroforming with constant fluid pressure. J. Mater. Process. Technol., 88: 51-56.

[11] Kandil A (2003) An experimental study of hydroforming deep drawing. J. Mater. Process. Technol. 134: 70-80.

[12] Zhang SH, Danckert J (1998) Development of hydro-mechanical deep drawing. J. Mater. Process. Technol. 83: 14 – 25.

[13] Tirosh J, Shirizly A, Ben-David D, Stanger S (2000) Hydro-rim deep-drawing processes of hardening and rate-sensitive materials. Int. J. Mech. Sci. 42: 1049-1067.

[14] Groche P, Metz C (2006) Investigation of active-elastic blank holder systems for high-pressure forming of metal sheets. Int. J. Mach. Tools Manufact. 46: 1271–1275.

[15] Groche P, Metz C (2005) Hydroforming of unwelded metal sheet using active-elastic tools. J. Mater. Proc. Tech. 168: 195-201

[16] Lang L, Danckert J, Nielsen KB (2004) Investigation into hydrodynamic deep drawing assisted by radial pressure Part I. Experimental observations of the forming process of aluminum alloy. J Mater Process Technol 148:119-131.

[17] Gorji A, Alavi-Hashemi A, Bakhshi-Jooybari M, Norouzi S, Hossinipour SJ (2011) Investigation of hydrodynamic deep drawing for conical-cylindrical cups. Int. J. of Advanced Manufacturing Technology. 56: 915-927.

[18] Lange K, (1985) Handbook of Metal Forming. McGraw-Hill Book Company, New York.

[19] Kawka M, Olejnik L, Rosochowski A, Sunaga H, Makinouchi A (2001) Simulation of wrinkling in sheet metal forming. J Mater Process Technol 109:283-289.

[20] Wong CC, Dean TA, Lin J (2003) A review of spinning, shear forming and flow forming processes. Int J Mach Tools Manufact 43:1419-1435.

[21] Liaghat G.H, Darvizeh A, Javabvar D, Abdollah A (2002) Analysis of explosive forming of conical cups, comparison of experimental and FEM simulation results. Amirkabir Journal 50:250-264 (in Persian).

Chapter 4

Forming of Sandwich Sheets Considering Changing Damping Properties

Bernd Engel and Johannes Buhl

Additional information is available at the end of the chapter

1. Introduction

1.1. Sandwich sheets

For many applications there are claims concerning the metal sheet, which cannot be reached with a single sheet. Composite materials, which can be classified in fiber-composites, sandwich materials and particle-composites are increasing year by year [1]. Sandwich respectively laminates with various thicknesses and materials of the different layers offer highly useful properties. As seen in Figure 1 every material has different mechanical, acoustical, tribological, thermal, electrical, chemical as well as environmental and technological properties.

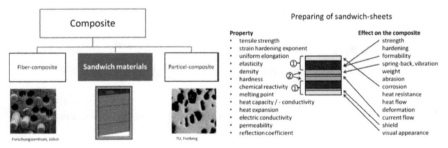

Figure 1. Classification of composites; preparing of sandwich sheets

Figure 1 illustrates an example of a fictive constellation of a composite. The price of the raw material itself has to be lower than the price of stainless steel. The nearly symmetrical sandwich should be resistant against corrosion and appear like stainless steel. For application in the automotive industry, it should be highly formable and reduce vibrations.

So the thin outer sheets are chosen of 1.4301, the supporting layer of DC06 and the vibration damping layer of a viscoelastic adhesive. At first, the metal layers of both sides are cladded 1. In step no. 2 they are bonded with a very thin adhesive layer (see chapter 5.4). To achieve advantages concerning the forming process (see chapter 4.1), a metal interlayer is included.

1.2. Failure-modes of sandwich sheets

Sheets with vibration damping qualities can be made of steel sheets enclosing a viscoelastic plasticcorelayer (see Figure 17). [2] The vibration energy of the oscillating coversheet is converted into heat.[3] [4] For automotive lightweight constructions, sandwich sheets with a noise-absorbent behavior are highly useful in the engine bay. [5]

Parts of sandwich sheets with viscoelastic layers excel with a higher security against cracking. [6]

Due to the relocatability of the outer and the inner sheet, leakage of e.g. an oil pan in contact with the subsoil or in case of a crash is unlikely for this sandwich in contrast to a comparable single layer sheet. The cover-sheets remain undamaged, but the bond can fail during forming. Under load, during and after the forming process special effects and failures occur. [7]

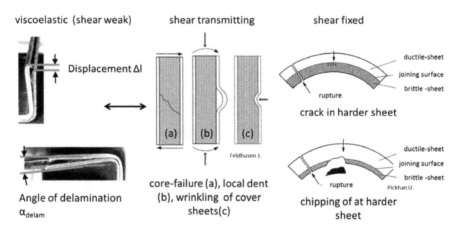

Figure 2. Failure modes of during forming of sandwich sheets (viscoelastic, shear transmitting and shear fixed)

Mainly displacement and delamination of the cover sheets occur as failure modes (see Figure 2). Wrinkling at the inside surface is detected especially by vibration-damping composite sheets with long non-deformed legs and small thickness of the inner layer. Plastics and adhesives creep under load. [8] Because of the residual stresses in the cover sheets or due to temperature influences, the composite delaminates often after hours or days, respectively.[9] To decay noises effectively, the viscoelastic interlayer has to be as thin as possible (chapter 5). In contrast to this, lightweight sandwich sheets achieve great stiffnesses with thick synthetic cores. [10] As well as by shear weak connection, the cover layer of shear transmitting sandwiches usually does not crack. Now, additive failures like core-failure, local dents and wrinkling of cover sheets limit the forming capabilities. [11] Typically, laminates which are jointed shear fixed crack on the brittle side. [12]

1.3. Classification of sandwich sheets

The joining process of several layers has a great influence of the formability and the damping behavior of the sandwich. Layers can be connected viscoelasticly by using an adhesive film. The cover sheets are allowed to slide on each other. For that displacement only a negligible shear force is necessary. This viscoelastic bond is weak. Shear transmitting bond lines like 1 or 2 component-adhesives are ductile. Parameters of the roll bonding process and surface treatments are shown in [10]. Layers which are connected shear fixed (cladded) can't slide on each other. [15] In a forming process the sheet behaves like a single sheet, but after forming, astonishing effects due to different Young's moduli and hardness occur.

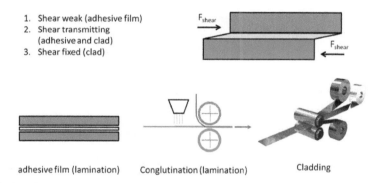

Figure 3. Classification of joining with respect to transmitting shear force

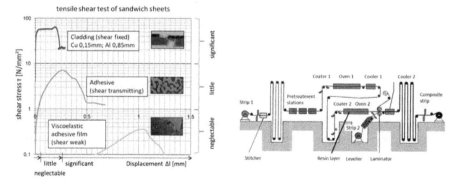

Figure 4. Tensile shear test of sandwich sheets; Strip production line similar to [3]

This classification is confirmed with tensile shear tests. As seen in Figure 4 the link surface of cladded sheets remains undamaged. This example shows a layer of copper (thickness $t_{co} = 0,15\,mm$) and aluminum (thickness $t_{al} = 0,85\,mm$). Under a shear stress of approximately $\tau = 60\,N/mm^2$, the layer of aluminum cracks under $\alpha = 45°$ degree. Cladded sandwich sheets are produced for thermal protection shields, pipe clamps, and safety components in automotive section, microwaves, cutting punches, heat exchanger and bipolar collector plate. [13],[14],[15]

Little shear stress can be transmitted with an adhesive layer which shows tow fracture appearances, the adhesive and cohesive crack. This type of sandwich sheets is often laminated as seen in Figure 4. A strip production line comprises a decoiler for the strip 1, an alkaline degreasing and chemical pretreatment of strip 1, a coater of strip 1 with the resin layer (applied on coater 2), a continuous curing oven 2, and a decoiler for strip 2. Then, it follows the laminator which rolls strip 1 and 2 into the composite strip material. After laminating the sandwich sheet is cooled in station 2 and recoiled. [5] Further this shear transmitting sandwich sheets are considered.

When viscoelastic interlayers, performed as an adhesive film, reach their maximal displacement, the adhesive detaches of the cover sheets.

2. Experimental tests and their simulation

Material properties will be determined by experimental tests. The connection should be classified with tensile shear tests. All tests are carried out at room temperature. For vibrations damping sandwich sheets, the parameters of different approaches will be computed. The advantages of the different approaches will be discussed.

Main experimental tests to get information of the forming behavior will be described and illustrated with examples.

2.1. Uniaxial tensile test

The relation between stress and strain for uniaxial tension is determined with tension tests. [16] The tests are carried out under an ambient air temperature of $t_{air} = 22\,°C$ with a tension velocity of $v = 0,05\,mm/s$. The yield strength ($R_{p0,2}$), tensile strength (R_m), and uniform elongation (φ_{gl}) of cover sheets were read out. DC04, DC06, 1.4301 and 1.4640 are chosen for the cover-sheet material. Three tensile tests carried out under the same conditions are shown in Figure 5 (cover sheet material: DC06).

For further analytical calculations of the strain-distribution, the bending-moment and the shear-force, the plastic behavior of the sheet material is described linearly with the gradient m, see equation (1).

$$k_{f\_lin} = R_{p0,2} + \varphi m \qquad m = \frac{(1+\varepsilon_{gl})R_m - R_{p0,2}}{\varepsilon_{gl} - \frac{R_{p0,2}}{E}} \qquad (1)$$

Out of a composition of common different material descriptions (Table 1), the stress-strain relation for numerical calculations is described according to Swift/Krupkowski k_{f_Swift} [17]. This three-parameter model describes an exponential function. The parameters can be found by using numerical optimizations. Furthermore, the parameters are determined using the instability-criterion, the strain hardening coefficient n, uniform elongation A_g and so on according [18]:

$$k_{f\_Swift} = b(c + \ln(1 + \epsilon))^d \qquad n = \varphi_v \approx \ln(1 + A_g) \qquad (2)$$

$$c = \sqrt[n]{\frac{R_{p0,2}}{R_m}} \frac{n}{e} \qquad d = (c + n) \qquad b = R_m * \frac{e^n}{d^d} \qquad (3)$$

So, the following Table 1 can be established.

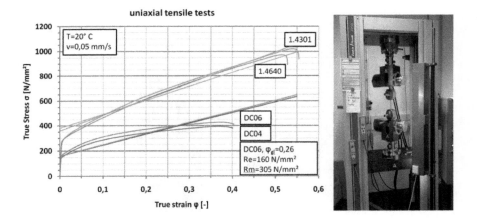

Figure 5. Uniaxial tensile tests of the metal sheet layer [6], Universal testing machine

	DC 04	1.4301	1.4640	DC06
d	0,284	0,443	0,370	0,236
c	0,061	0,043	0,029	0,005
b	545	1337	1268	540
n	0,2231	0,4007	0,3414	0,2311

Reihle/Nadai:	$k_f = c_1 * \varphi_v{}^n$	(4)
Swift/Kurp.:	$k_f = c_1(c_2 + \varphi_v)^n$	(5)
Gosh:	$k_f = c_1(c_2 * \varphi_v)^n - c_3$	(6)
Voce:	$k_f = c_2 - (c_2 - c_1)e^{(-c_3\varphi_v)}$	(7)
Hocket-Sherby:	$k_f = c_2 - (c_2 - c_1)e^{(-c_3*\varphi_v{}^n)}$	(8)
Extrapolation Rein-Alu :	$k_f = c_1 * \varphi_v{}^n * e^{\left(\frac{-c_3}{\varphi_v}\right)}$	(9)

Table 1. Material parameters for Swift/Krupkowski, composition of common material models

2.2. Tensile shear test

During the forming process the cover layers of the sandwich slide on each other. By shifting the cover plates the adhesive layer is sheared. Tensile shear tests provide the stress-displacement behavior of the adhesive for a constant shear rate at room temperature. They can be transferred directly to the forming process [6]. Different adhesives applied in liquid state were investigated. In addition, sandwich sheets of Bondal® [5] CB (Car Body) from ThyssenKrupp are considered. As seen in Figure 6, the specimens were prepared according to DIN 53 281 [19] respectively [20] but also [21]. The thicknesses of the cover sheets ($s = 0,75\ mm$) and of the adhesive layer ($s_{ad} = 0,05\ mm$) are given by the composite material. In Figure 6 three tensile shear tests of sandwich sheets with different testing velocities are shown. The shear stress is calculated from the initial overlapping length $L_{ü}$ and the gauged force F.

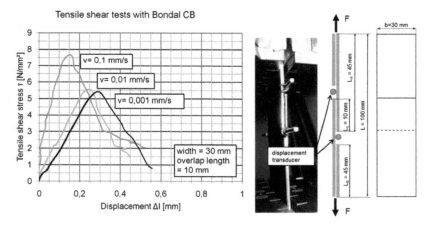

Figure 6. Tensile shear test with Bondal CB (cover sheets of DC06), specimen geometry

The adhesive layer fails by reaching the maximum force. For composites with a shear transmitting adhesive a shear stroke of $\overline{\Delta l}_{max} \approx 0,22\ mm$ was determined at a maximum shear stress of $\overline{\tau}_{max} \approx 5,5\ \frac{N}{mm^2}$ (see Figure 6). These values were computed by arithmetic mean. With higher velocity, the shear stress increase, but the tolerable displacement drops. The specific stiffness of the adhesive ($\frac{E}{K_{nn}} = \frac{\tau_{max}}{\Delta s}$), a measurement to describe the increase of the shear stress, rises.

Delamination and displacement of laminated metal sheet can be analyzed by using finite elements. In Figure 7 the experimental and numerical tensile shear tests of the adhesive film are shown. In literature, several investigations determined the stress distribution depending on the overlapping lengths [8]. As the experimental tensile shear test, the force F and displacement Δl are gauged with numerical calculations. The shear stress is determined by the force, as well as by the initial overlapping length. There are different modes of delamination. G. Alfano and M. A. Crisfield e.g. established an interaction model "mixed-mode" which summarizes many fracture criteria proposed in literature. [22] To describe the elasticity of the adhesive, the traction/relative displacement law is used. The damage law with nominal stress of the normal mode and the both other directions is chosen. Numerical results of the uniaxial tensile shear tests fit to the experiments. Special attention for forming sandwich sheets is laid on the failure of the adhesive bond. As seen in Figure 7 the elastic and damage behavior is calculated exactly, but the maximal shear force and displacement depart up to 9 %.

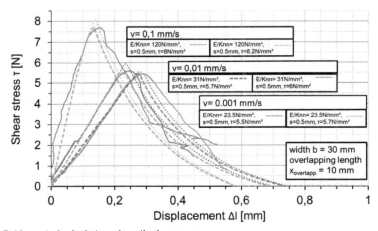

Figure 7. Numerical calculation of tensile shear tests

3. Forming tests: Die-bending

To predict the suitability of vibration-damping sandwich sheets for forming, material properties and interactions are calibrated with the simulation of the die-bending tests. V-die bending with different temperatures are considered in [23] and of velocity in [24]. Therefore specimens with a special grid on the surface of edge were prepared.

Optimized adhesive parameters for numerical calculation are applied on the shear test (see Figure 7). With bilinear, quadrilateral elements the material is performed. Reduced integration plane stress space with hourglass control (CPS4R) is used. Hard pressure-overclosure behavior is approximated with the penalty method. The penetration distance is proportional to the contact force. Contact is implemented frictionless to assume that surfaces in contact slide freely and isotropic with a friction-coefficient of $\mu_{werk} = 0,1$.

In Figure 8 the v-die bending process of a symmetrical sandwich with a length of 2L is considered. The stress, strain and displacement of both cover sheets out of DC06 with a thickness of s = 0,7 mm are determinant for two zones, the bending area and the adjacent area. By the punch radius r and the bending angle α, the bending area is defined. In contrary to homogeneous sheets, the adjacent area of the bending area shows a significant stress distribution. This leads to a plasticization of the primarily adjacent area. The residual stress also leads to delamination even after weeks or even under influence of temperature.

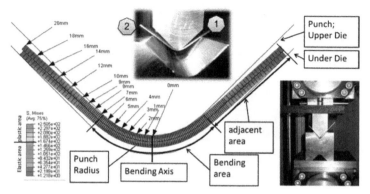

Figure 8. Pilot project: Experiment v-die bending, v-die bending with cover sheet material DC06

The adhesive layer transmits shear-force of the bending area into the adjacent area. For distances $t = [1, 2, 3...]$ mm from the bending axis the displacement is measured in the simulation and experimental tests. Accurate preparing of the specimens for die bending has a great influence on applicability of the results. After polishing the surfaces of the edges, a micro grid is scribed with a depth of $t_{dia} = 3$ μm. At the inflection line of the chamfer with the width $b_{dia} = 5$ μm the light reflects and a sharp line can be seen. For scribing, a diamond with a point angle of $\alpha_{dia} = 130°$ is used. Figure 9 shows calculated and experimental displacements of cover sheets after bending. With the influence of manufacturing, the enlarged displacements of experimental tests are explained. The simulation-model is verified.

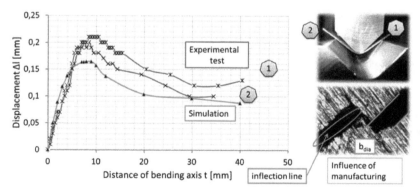

Figure 9. Pilot project v-die bending: calculated and experimental displacement of cover sheets, material DC06 (b)

Normally, only the edges of a formed part can be seen. Thus, the displacement of the sheets in the bending area is not detectable. At first the displacement of the edges depends on the side length L = [40, 35, 30, and 25] mm, see Figure 10. As expected, increasing side length causes decreasing displacement.

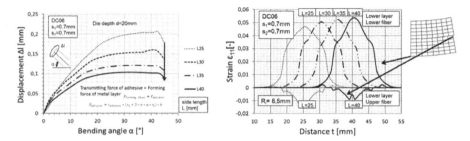

Figure 10. Numerical results of die-bending; displacement Δl over bending radius α for variation of side length L (a), elongation of the external fiber of the both layers with bending axis at distance t = 40 mm (b)

Also the strain of the upper and under fiber of the under layer in bending direction is shown. With larger side lengths the strain increases and a special peek of strain occurs in the bending center. The neutral axis of the lower cover-layer moved depending on the side length in direction to the upper fiber. In the upper layer, the neutral axis moved to the midst of the sandwich sheet, too. Further strain distributions of other fibers, the influence of friction and several displacements dependent on the distance from the bending axis in the bending and adjacent area are shown in [25]. Also the influence of adhesive in contrast to tow single sheets, which slide frictionless on each other, can be seen. So a cutback of the edge displacement from Δl = 0,37 mm to 0,09 mm (L = 40 mm) is achieved with a viscoelastic adhesive. From this it follows that the shear transmitting interlayer doesn't fail at the edges. But at the beginning of the adjacent area, the displacement crosses Δl = 0,15 mm by already a bending angle α = 30°.

Even sandwich sheets with large side lengths of for example L = 40 mm can gain an inner failures during forming. Simulations with adjusted layer-thickness are shown in chapter 6.

4. Plastomechanical preliminary design

Correct use of simulation tools demands experiences in element types, material laws, contact properties, solver and integration step characteristics to evaluate numerical results. Moreover the calculation requires high calculation time and performance. [26]

To reduce the calculation time a plastomechanical preliminary design is established (see Figure 11). Several numerical calculations describe the bending behavior of three-layer sandwich sheets with viscoelastic interlayers. Numerical investigations are made by M.

Weiss for the elastic [27] and TAKIGUCHI for the plastic [28] bending process of three layer sandwich plates.

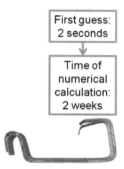

Figure 11. Aim of plastomechanical preliminary design

Proposals to calculate the reached length of shear fixed sandwich sheets are established of Hudayari [29]. With the following two methods "viscoelastic" and "shear transmitting", simple proposals for designing the thickness ratio of the metal layers and the required shear stress are made. Especially the "viscoelastic" description of the forming of sandwich sheets is qualified for basic statements.

4.1. Three-layer sandwich with viscoelastic interlayer

Sandwich sheets with an adhesive film show displacements and delamination very clearly. At a specific displacement $\overline{\Delta l}_{max}$ the interlayer fails. When the adhesive delaminates from the sheets, the cohesive force is small and can be neglected. Therefore, the cohesive force is not include in this "viscoelastic model"; neither to determine displacement, nor of the angle of delamination α_{delam}.

As explained in [6], the Euler–Bernoulli beam theory [30] is used. The neutral axis remains in the middle of each layer. With following geometric relations, the technical strain ε can be determined in dependency of the radius of the neutral axis R_n and the control variable y:

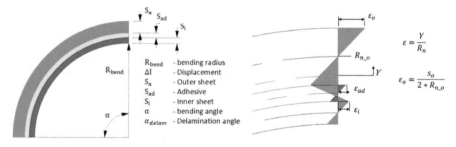

Figure 12. Geometrical description of the bending process [6] and strain distribution over thickness

$$\Delta l = \alpha * 0.5 * (s_a + s_i) \tag{10}$$

Independent of the bending radius R_{bend}, the geometric displacement Δl is determined with the bending angle α and the sum of layer-thicknesses S_a and S_i (equation (10)). This approach fits quite well [6]. Since the adhesive layer is thin, it can be neglected. Over the thickness of the cover-layers, as well as over the bending angle, a linear strain distribution is assumed. The spring-back angle α_{sb} is calculated from the integration of the back-curvature under usage of a linear-plastic material model according [31]:

$$\alpha_r = \int_{x=0}^{x=r\alpha} \frac{2*\sigma_{ael}}{E*s} dx = \frac{2*\sigma_{ael}}{E*s} R_n \alpha \tag{11}$$

With σ_{a_el} for the stress which relaxes during spring back:

$$\sigma_{a\_el} = \frac{6R_n^2}{s^2} \left(\frac{2}{3} \left(\frac{R_{p0,2}}{E} \right)^3 \left(E + \frac{m}{2} \right) - \left(\frac{R_{p0,2}}{E} \right)^2 R_{p0,2} \right) + \frac{s}{2r} m + 1,5 (R_{p0,2} - \frac{R_{p0,2}}{E} m) \tag{12}$$

Neglecting small terms, the spring-back angles α_{sb} are determined for each layer:

$$\alpha_{sb} \approx 3 \frac{R_{p0,2}}{E} * \frac{R_n}{s} * \alpha \tag{13}$$

The angle of delamination α_d from the difference of the two spring-back angles is calculated:

$$\alpha_d = \alpha_{sb\_a} - \alpha_{sb_i} \tag{14}$$

For the same materials of both cover layers the delamination α_d is established relating to the inner radius of the sandwich sheet R_{bend}:

$$\alpha_{d\_mat} \approx 3\alpha \frac{R_{p0,2}}{E} \left(\frac{s_i R_{bend} + s_i s_i + s_{ad} s_i - s_a R_{bend}}{s_i * s_a} \right) \tag{15}$$

And for the same thicknesses $S = S_i = S_a$ of the cover sheets:

$$\alpha_{d\_th} \approx \frac{3\alpha}{s} \left(\frac{R_{p0,2\_a}}{E_a} * (R_{bend} + s_{ad} + 1,5 * s) - \frac{R_{p0,2_i}}{E_i} (R_{bend} + 0,5 * s) \right) \tag{16}$$

Standardly, sandwich sheets are produced and as a matter of fact have the same cover-materials and thicknesses. S_{ad} is the thickness of the adhesive:

$$\alpha_{d\_mat\_th} \approx 3\alpha \frac{R_{p0,2}}{E} \left(\frac{s^2 + s_{ad} s}{s^2} \right) \tag{17}$$

Regarding to these calculations the following options to prevent the tendency of delamination were identified by [6]:

- The softer material should be chosen as the inner layer. But the inner layer has to bear the forming forces.
- Angle increases by reducing the thickness. If the thinner layer forms the inner curve, the inner layer springs against the outer. [6] shows, that no delamination occurs for a specific thickness ratio.
- Overbending is another method to avoid delamination. So elastic stress, which results from the different angle, leads to a compressive stress at the contact surfaces after bending back.

- The delamination can be prevented by pre-bending with a smaller radius. With the expansion of the radius in the forming process, the different angle of the cover sheets is used.

4.2. Sandwich sheets with shear transmitting interlayer

As described by the tensile shear tests in chapter 2.2, the metal layers can slide on each other under shear force. The adhesive layer has been considered as viscoelastic. Now, sandwich sheets with shear transmitting interlayers are considered. The neutral axis of the cover lower layer moved depended on the side length (chapter 3) and the shear stress τ_{max} of the adhesive. In the upper layer, the neutral axis moved to the midst of the sandwich sheet, too.

So the shear force of the adhesive, which superposes the bending moment M_b with the force $F_{\ddot{u}}$, constitutes the shifting of the axis. The shifting of the neutral axis is expressed by the k-factor (Figure 13). The compressive strain and elongation of the cover sheets changes. Part (b) of Figure 13 shows the outer cover sheet of a symmetrical shear-fixed sandwich. Strain modes of a sandwich layer, which transmits no $F_{\ddot{u}} = 0$, $k_i = k_a = 0$ (c), tension (a, b) or even compression forces (d, e), are shown.

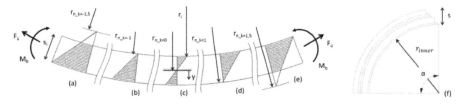

Figure 13. Principle of k-factor for each layer (a-e), geometric of a sandwich with different layer-thicknesses (f) and a layer under a bending moment (g)

The strain distribution of sandwiches with i metal layers, which may have different thicknesses s_i, can be described by the k-factor. For each metal layer with the thickness s_i, the neutral axis R_n i is computed dependent on the thickness of the sandwich s:

$$r_{m,i,n} = r_{inner,i,n} + \left(1 + k_{i,n}\right) \cdot \frac{s_i}{2} \qquad (18)$$

The bending radius r_{inner} refers to the inside of the sandwich. Considering this, the true bending moment $M_{B,i,n}$ is:

$$M_{B,i,n} = \underbrace{\int_{0}^{\frac{s_i}{2}\left(1-k_{i,n}\right)} \sigma_{plast,i,n} \cdot b \cdot y \cdot dy}_{tension} + \underbrace{\int_{-\frac{s_i}{2}\left(1+k_{i,n}\right)}^{0} \sigma_{elast-plast,i,n} \cdot b \cdot y \cdot dy}_{compression} \qquad (19)$$

Integration of the bending moment leads to dependency of the k-factor:

$$M_{B,i,n} = A_i \left(A_{0,i,n} + A_{1,i,n} \cdot k_{i,n} + A_{2,i} \cdot k_{i,n}^2 \right) + m_i \cdot I_{Z,i} \frac{1 + 3 \cdot k_{i,n}^2}{r_{m,i,n}} \tag{20}$$

$$A_i = R_{p0.2,i} \cdot \left(1 - \frac{m_i}{E_i} \right) \cdot b$$

$$A_{0,i,n} = \frac{s_i^2}{4} - \frac{1}{3} \cdot \left(\frac{R_{p0.2,i}}{E_i} \right)^2 \cdot \left(r_{innen,i,n} + \frac{s_i}{2} \right)^2$$

$$A_{1,i,n} = -\frac{1}{3} \cdot \left(\frac{R_{p0.2,i}}{E_i} \right)^2 \cdot \left(r_{innen,i,n} + \frac{s_i}{2} \right) \cdot s_i \tag{21}$$

$$A_{2,i} = \frac{s_i^2}{4} - \frac{1}{3} \cdot \left(\frac{R_{p0.2,i}}{E_i} \cdot \frac{s_i}{2} \right)^2$$

With the moment of inertia $I_{z,i}$ of a rectangular cross-section:

$$I_{z,i} = \frac{b \cdot s_i^3}{12} \tag{22}$$

Substituting the bending moment under load M_B with the elastic bending moment $M_{B\text{-}el}$, the radius of curvature K can be determined for the unloaded case.

$$M_{B-el,i,n} = \frac{b}{r_{R,i,n}} \cdot E_i \cdot \int_{-\frac{s_i}{2}(1+k_{i,n})}^{\frac{s_i}{2}(1-k_{i,n})} y^2 \cdot dy = \frac{1}{r_{R,i,n}} \cdot E_i \cdot I_{z,i} \cdot \left(1 + 3 \cdot k_{i,n}^2 \right) \tag{23}$$

$$\kappa_{R,i,n} = \frac{M_{B,i,n}}{E_i \cdot I_{z,i} \cdot \left(1 + 3 \cdot k_{i,n}^2 \right)} \tag{24}$$

So, the remaining radius α_{bl} is calculated:

$$\alpha_{bl,i,n} = \int_0^{r_{m,i,n} \cdot \alpha_n} \kappa_{bl,i,n} \cdot d\alpha = \left(1 - \kappa_{R,i,n} \cdot r_{m,i,n} \right) \cdot \alpha_n \tag{25}$$

$$\alpha_{bl} = 1 - B \frac{\left(B_0 + B_1 \cdot k + B_2 \cdot k^2 + B_3 \cdot k^3 \right)}{\left(1 + 3 \cdot k^2 \right)} - \frac{m}{E} \tag{26}$$

$$B = \frac{R_{p0.2}}{E \cdot I_Z} \cdot \left(1 - \frac{m}{E}\right) \cdot b$$

$$B_0 = \frac{s^2}{4} \cdot \left(r_{innen} + \frac{s}{2}\right) - \frac{1}{3} \cdot \frac{R_{p0.2}^2}{E^2} \cdot \left(r_{innen} + \frac{s}{2}\right)^3$$

$$B_1 = \frac{s^3}{8} - \frac{1}{2} \cdot \frac{R_{p0.2}^2}{E^2} \cdot \left(r_{innen} + \frac{s}{2}\right)^2 \cdot s \qquad (27)$$

$$B_2 = \left(1 - \frac{1}{3} \cdot \frac{R_{p0.2}^2}{E^2}\right) \cdot \frac{s^2}{4} \cdot \left(r_{innen} + \frac{s}{2}\right)$$

$$B_3 = \left(1 - \frac{1}{3} \cdot \frac{R_{p0.2}^2}{E^2}\right) \cdot \frac{s^3}{8}$$

$$\Delta l_{j,n} = \Delta l_{j,n-1} + \left(\alpha_{bl,i,n} - \alpha_{bl,i-1,n}\right) \cdot r_{m,n} - \left(1 + k_i\right) \cdot \frac{s_i}{2} \cdot \alpha_{bl,i,n} - \left(1 - k_{i-1}\right) \cdot \frac{s_{i-1}}{2} \cdot \alpha_{bl,i-1,n} \quad (28)$$

$$\Delta \alpha_{j,n} = \Delta \alpha_{j,n-1} + \left(\kappa_{R,i-1,n} - \kappa_{R,i,n}\right) \cdot r_{m,n} \cdot \alpha_n \qquad (29)$$

In a multilayer sandwich sheet, the displacement Δl and the angle of delamination $\Delta \alpha$ between two layers can be calculated as shown in equation (28), (29). The determination of the k-factor and the following proposal list to avoid or minimize the failure modes at the edges are according to [25] :

- Keep the required overlapping length, respectively the side length
- Choose an adhesive depending on the shear force which is calculated
- Enlarge the number of metal layers
- Reduce the thicknesses of the outer layers
- Choose a material for outer layers with low yield strength
- Choose a small hardening coefficient

5. Acoustical calculations

5.1. The loss factor, a measurement for damping behavior of a material

The transmission of structure-borne noise depends highly on the behavior of a material. Through the temporal offset of the shear stress and strain, the vibration energy is converted into heat energy. [32], [33], [34]

A gauge of the internal damping of a material, the absorption capacity of vibrations, is the loss factor tan δ. With increasing loss factor the material behavior approaches a Newtonian fluid with viscosity. Even if the metal sheet is not damped by a polymeric coating or interlayer, the vibration of the sheet decays after a certain time. This effect is due to internal friction of the solid. Compared to synthetic materials, steel converts a lower amount of vibration energy per oscillation into heat. The decay will take more time compared to a sandwich sheet. Using the poisson's ratio ν, the shear modulus is determined according to:

$$G = \frac{E}{2 \cdot (1 + \vartheta)} \tag{30}$$

The loss factor is the ratio of storage modulus G' and loss modulus G'' [3]:

$$tan\ \delta = \frac{G'}{G''} \tag{31}$$

Complex shear modulus $G^* = G' + j \cdot G''$ with its amount $|G^*| = \sqrt{(G')^2 + (G'')^2}$ is highly dependent on temperature and amplitude. Often in a range of about 60 °C a constant value is assumed.

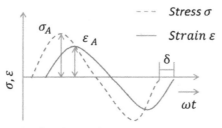

Figure 14. Offset between stress and strain over time

Material	loss factor tan δ
Steel	0,0006 – 0,0001
Aluminum	0,0001 – 0,001
Gray iron	0,01 – 0,02
Film of Bitumen	0,2 – 0,4
Damping mat	0,2 - 1

Material	tan δ [-]	E-Modulo [N/mm²]	T_{amb} [°C]	f [1/s]
Polyvinyl chloride	1,8	30	92	20
Polystyrene	2,0	300	140	2000
Polyisobutylene	2,0	6	20	3000
nitrile rubber	0,8	330	20	1000
hard rubber	1,0	200	60	40
Polyvinyl chloride with 30% plasticizer	0,8	20	50	100

Table 2. Loss factor tan δ for different materials under ambient temperature of T_{amb} = 20 °C [34] and Material properties according to [33]

Figure 15 shows the relation of the loss factor tan δ and the shear modulus, which leads to following equation:

$$G' = G \cdot cos(\delta) \tag{32}$$

Thus, the storage modulus G' can be detremined as:

$$G'_2 = G \cdot cos(arctan(tan\delta_2)) \tag{33}$$

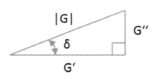

Calculated shear modulus G	
Young's modulus adhesive E [N/mm²]	6
Poisson's number ν [-]	0,49
Shear modulus G* [N/mm²]	2,013422819
Storage modulus G' [N/mm²]	0,900430058

Figure 15. Representation of a complex shear modulus |G*|, calculated values for the adhesive

5.2. Determination of decay behavior

In most calculations the damping effect is based on a linear combination of mass M and stiffness K. The sandwich sheet with viscoelastic damping subject following elasto mechanical system:

$$f_{(t)} = M\ddot{y} + C\dot{y} + Ky \tag{34}$$

M	=	mass matrix
K	=	stiffness matrix
C	=	damping matrix
y	=	displacement vector
ẏ	=	velocity vector
ÿ	=	acceleration vector

Table 3. Parameters of the elasto-mechanical system: sandwich sheet with viscoelastic interlayer

The equation of motion in time includes damping by energy dispersion. Simplified, an initial displacement is assumed to generate the oscillation. This means that no force or impulse is considered in the calculation. This assumption simplifies the determination of the initial conditions to:

$$0 = \ddot{y} + \frac{c}{m}\dot{y} + \frac{k}{m}y \tag{35}$$

The terms $\frac{k}{m}$ and $\frac{c}{m}$ of the linear homogeneous equation can be transcribed with the relation of angular frequency of the undamped harmonic oscillation ω_0^2 and the damping ratio D to get the differential equation:

$$\omega_0^2 = \frac{k}{m} \tag{36}$$

$$D\omega_0 = \frac{c}{m} \tag{37}$$

$$0 = \ddot{y} + D\omega_0\dot{y} + \omega_0^2 y \tag{38}$$

The Euler representation with the function $sin(\omega_d \cdot t + \varphi_0)$ is used with the damped angular frequency ω_d. From experimental tests it is well known, that the decay based on displacement behaves $\hat{y}_0 \cdot e^{-\delta \cdot t}$ [35].

With $\hat{y}_0$, the value of the displacement at time $t = 0$:

$$y_{(t)} = \hat{y}_0 \cdot e^{-D\omega_0 \cdot t} \cdot sin(\omega_d \cdot t + \varphi_0) \tag{39}$$

With the following characteristics of:

Damping/ Decay constant:

$$\delta = tan\delta_{Sandwich} \cdot \pi \cdot f \tag{40}$$

Resonance quality:

$$Q = \frac{1}{tan\delta_{Sandwich}} \tag{41}$$

Damping ratio:

$$D' = \frac{tan\delta_{Sandwich}}{2} \tag{42}$$

The natural angular frequency of damped oscillation ω_d is calculated of angular frequency ω_0 and damping constant δ.

$$\omega_d = \sqrt{\omega_0^2 - \delta^2} \tag{43}$$

Due to the viscous damping, only the physically significant context $\omega_0^2 - \delta^2 > 0$ is considered further. With the dimensionless damping ratio D, the decay and input angular frequency can be compared:

$$D = \frac{\delta}{\omega_0} \tag{44}$$

5.3. Damping-behavior of a three layer sandwich

Ross, Kerwin and Unger (1959) [36] calculated with their approach "Damping Model" the damping behavior of a three-layer composite. Many modification has been made [37], [38], [39] with are summarized in [40]. Starting with the elastic bending moment M of the three-layer sheet, the shear forces and shear strains are calculated:

$$M = B\frac{\partial \Phi}{\partial \chi} = \Sigma_1^3 M_{ii} + \Sigma_1^3 F_i H_{io} \tag{45}$$

Where B is the flexural rigidity per unit width of the composite plate, M_{ii} the Moment of exerted by the forces on the ith layer about its own neutral plane, F_i the net extensional force on the layer and H_{io} the distance from the center of the ith layer to the neutral plane of the composite beam. [33] predicted the loss of "thick plates with a thin intermediate layer":

$$tan\delta_{Sandwich} = tan\delta_2 \cdot \frac{h \cdot g}{[[1+(1+i \cdot tan\delta_2) \cdot g]^2 + g \cdot h \cdot [1+g \cdot (1+tan\delta_2{}^2)]]} \tag{46}$$

With the "geometric parameter" 1/h, the distance between the neutral fiber of the sandwich structure and the coversheets is described.

$$\frac{1}{h} = \frac{B_1' + B_3'}{a^2} \cdot \left(\frac{1}{E_1 \cdot d_1} + \frac{1}{E_3 \cdot d_3}\right) \tag{47}$$

And a „shear-parameter" g:

$$g = \frac{G_2'}{d_2 \cdot k^2} \cdot \left(\frac{1}{E_1 \cdot d_1} + \frac{1}{E_3 \cdot d_3}\right) \tag{48}$$

Parameter "a" is the distance of the neutral fibers of the cover plates:

$$a \approx d_2 + \frac{(d_1 + d_3)}{2} \tag{49}$$

And the number of wave k:

$$k = \left(\frac{\omega^2 \cdot m'}{B}\right)^{\frac{1}{4}} \tag{50}$$

With the mass per unit length m':

$$m' = \rho \cdot S \tag{51}$$

The cross-sectional area S:

$$S = lg(d_1 + d_2) \tag{52}$$

And with the specific total bending stiffness B':

$$B' = (B'_1 + B'_3) \cdot \left(1 + \frac{g \cdot h}{1 + g \cdot (1 + i \cdot tan\delta_2)}\right) \tag{53}$$

Wherein the specific bending stiffness B_i for a plane sheet layer can be calculated as:

$$B'_i = \frac{1}{12} \cdot E_i \cdot d_i^3 \tag{54}$$

5.4. Influence of forming on damping-behavior of three layer sandwich sheet

As seen in chapter 0 the bending stiffness B_i for a sheet layer is significant for decay-behavior. With increasing stiffness of the cover layers, the total loss factor of the sandwich sheet decreases. Accordingly, the forming geometry has a big influence on the damping behavior. Cover-sheets with a thickness of $d_1 = d_2 = 1\,mm$ and a width and length with $b = l = 30\,mm$ and the viscoelastic interlayer (calculated adhesive, Figure 15) are shown in Figure 19. No. 1 shows the decay curve of an unformed sheet in contrast to a sheet with bended edges and a v-profile. All three sheets have the same initial width b. A great influence of the bending stiffness can be seen.

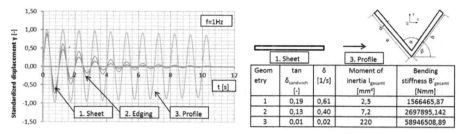

	1. Sheet			3. Profile	
Geometry	tan $\delta_{sandwich}$	δ	Moment of inertia I_{gesamt}	Bending stiffness B'_{gesamt}	
	[-]	[1/s]	[mm⁴]	[Nmm]	
1	0,19	0,61	2,5	1566465,87	
2	0,13	0,40	7,2	2697895,142	
3	0,01	0,02	220	58946508,89	

Figure 16. Damping effect of a three layer sandwich sheet, for an unformed, standardized displacement

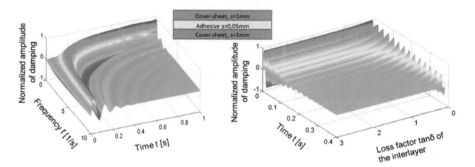

Figure 17. Influence of frequency and loss factor tanδ for a three layer sandwich-sheet

In Figure 17 the decay curve for different frequencies is shown. The loss factor tan δ of the adhesive is the main parameter of damping for a three layer sandwich-sheet. This factor is varied from zero to 3. A loss factor of tan δ = 2 is used for further investigations.

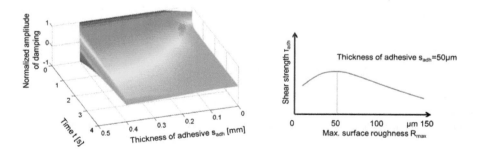

Figure 18. Influence of adhesive-thickness on damping behavior; and shear strength [8]

With smaller thicknesses of the adhesives, better damping behavior can be achieved. For a zero thickness, the adhesive transmits no bending waves. In Figure 18 the relationship between adhesive thickness, surface roughness R_{max} and bond strength is shown according to [8]. It is recommended that the adhesive thickness is equal to the surface roughness R_{max}. A smaller thickness avoids a complete coating. To achieve improvements in damping behavior the surface roughness and the thickness of the adhesive layer has to be reduced.

6. Optimization of vibration damping sandwich sheets

6.1. Comparison of numerical and plastomechanical calculations

Numerical die-bending results of side length L = 40 mm are shown in Figure 10. Now, the strain distribution of the upper and lower fiber of both layers are shown and compared with the plastomechanical preliminary design. Computation according to [25] offer a k-factor of $k_{i/o} = \pm 1$, which fits quite good for the maximum strain, calculated with a numerical method in the midst of the bending. The real positions of the neutral axis are shown schematically.

Inaccuracies of preliminary design are ascribed to the [25]:

*   Constant thickness during and after the bending process
*   Negligence of elastic behavior of the cover sheets
*   Linearization of the material properties
*   Linearization of the elongation distribution
*   Constant strain over bending angle

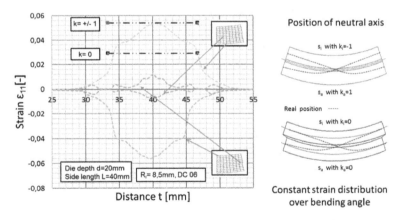

Figure 19. Numerical results of elongations of the external fibers (bending axis at distance t = 40 mm)

6.2. Forming tendencies of multilayer sandwich sheets

Thee founded results to influence the forming behavior, are further tested with simulation of v-die-bending. With a punch radius of r = 8,5 mm, a side length of L = 40 mm and a constant thickness of the sandwich s = 1,4 mm, the five typical constellations for different layer thicknesses (Figure 20) are calculated. The same material as in chapter 3 and 6.1 is used. For this calculation, the adhesive thickness is neglected.

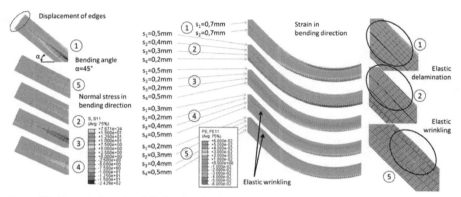

Figure 20. Numerical results of die-bending for five multilayer sandwich sheet

For this great side length, the displacement after forming is very small and no failure mode occurs for all five constellations. The symmetrical sandwich ($s_1 = s_2 = 0,7\ mm$) sheet no. 1 shows less spring back than all of the other constellations with 4 metal layers. The highest value for spring back shows no. 4 with increasing layer thickness. The remaining stress which can cause failure (chapter 1.2) decreases by using unsymmetrical thicknesses (no. 2, 3 and 4). As proposed in chapter 4.2, the number of layers influences the strain distribution. Minimal elongation shows specimen no. 3 and compression no. 5. Because of the minor stiffness no. 5 tends to buckling. This is an initial point for inner failures. The different spring back of each layer can be absorbed by the adhesive.

The normal plastic strain component in bending direction depends on the thickness of the metal layers. The thinner the inner cover sheet is, the minor is the plastic strain and the displacement of the edges (see no. 2-4). But a thin inner layer tends to buckling. To get the lowest normal stress in bending direction, the thickness should be increased as seen in configuration no. 4. Also the spring-back of the undamaged sandwich depends on the layer-configuration.

An example for an application of a commercial three-layer sandwich sheet in the automotive industry is shown in [25]. For this profile, formed by rolling the failures displacement, delamination and buckling could be predicted and verified with experimental tests.

7. Conclusions and forecast

At the Chair of Forming Technologies at the University of Siegen vibrations damping composite sheets were investigated regarding their forming limits.

During forming, failure modes like delamination, displacement and buckling of the cover sheets occur. The mechanical properties of the metal layers are determined by uniaxial tensile tests. Tensile shear tests are carried out under variation of the shear velocity. The numerical calculations are calibrated with the tensile shear test. To verify the adhesive description, maximal tension and displacement in normal direction will be proved.

A plastomechanical preliminary design has been developed for sandwich sheets with a viscoelastic and shear transmitting interlayer. Especially displacement and delamination are described with preliminary design and verified with numerical calculations. But this plastomechanical preliminary design shows improvable deliverables. Especially the inaccuracies of the strain distribution over thickness and bending angle should be improved. Therefore many experimental forming tests with different materials are planned.

Instead of symmetrical sandwiches, significant improvements can be achieved by e.g. reducing the thickness of the outer metal layer. With increasing the number of sheets the total thickness of the composite can be achieved. On the other hand thin inner layers tend to buckling. The buckling tendency should be investigated further.

Author details

Bernd Engel and Johannes Buhl*
University of Siegen, Chair of Forming Technology, Siegen, Germany

Acknowledgement

A part of this present Investigation has been financially supported of Bundesministerium für Wirtschaft und Technologie (BMWi).

8. References

[1]   Roos E, Maile K. Werkstoffkunde für Ingenieure Berlin Heidelberg: Springer; 2005.

[2]   Lange K. Umformtechnik: Blechbearbeitung, Handbuch für die Industrie und Wissenschaft Berlin Heidelberg: Springer-Verlag; 1990.

[3]   Jandel B, Meuthen AS. Coil Coating Wiesbaden: Vieweg & Sohn Verlag; 2008.

[4]   Antiphon. antiphon® MPMTM - vibration damping sandwich material. [Online].; 2010 [cited 2010 04 21. Available from: www.antiphon.se.

[5]   ThyssenKrupp Steel. Bondal® Körperschalldämpfender Verbundwerkstoff. [Online].; 2009 [cited 2010 04 21. Available from: www.thyssenkrupp-steel.com.

[6]   Engel B, Buhl J. Metal Forming of Vibration-Damping Composite Sheets. steel research international. 2011;(DOI: 10.1002/srin.201000205).

[7]   Hellinger V. New Possibilities for Improved Bending of Vibration Damping Laminated Sheets. CIRP annals: manufacturing technology. 1999.

[8]   Habenicht G. Kleben Heidelberg: Springer-Verlag; 2009.

---

* Corresponding Author

[9] Keßler L. Simulation der Umformung organisch beschichteter Feinbleche und Verbundwerkstoffe mit der FEM Aachen: Shaker; 1997.

[10] Palkowski H, Sokolova O, Carrado A. Reinforced matel/polymer/metal sandwich composite with improved properties. In The Minerals M&MS. TMS 2011, 140th Annual Meeting & Exhibition. New Jersey: Joh Wiley & Sons, Inc., Hoboken; 2011.

[11] Sirichai T. Modellierung und Simulation eines Verbunds von Sandwichplatten zur Entwicklung einer mechanischen Verbindungstechnik RWTH Aachen: Doctoral thesis; 2007.

[12] Pickhan U. Untersuchungen zum abrasiven Verschleißverhalten, Biegeumformen und Verbindungsschweißen walz- und schweißplattierter Grobbleche Siegen: Höpner u. Göttert; 1998.

[13] Yanagimoto J, Oya a T, Kawanishi a S, Tiesler N, Koseki T. Enhancement of bending formability of brittle sheet metal in multilayer metallic sheets. CIRP Annals - Manufacturing Technology 59. 2010: p. 287–290.

[14] Ohashi Y, Wolfenstine J, Koch R, Sherby O. Fracture Behavior of a Laminated Steel–Brass Composite in Bed Tests. Materials Science and Engineering. 1992.

[15] Köhler M. Plattiertes Stahlblech Düsseldorf: Stahl-Informations-Zentrum; 2006.

[16] Deutsches Institut für Normung. DIN 50125 Prüfung metallischer Werkstoffe - Zugproben. In. Berlin: Beuth Verlag GmbH; 2004.

[17] Swift HW. Plastic Instability under Plane Stress. Journal of the Mechanics and Physics of Solids 1, Department of Engineering, University of Sheffield UK. 1952: p. 1–18.

[18] Groche P, von Breitenbach G, Steinheimer R. Properties of Tubular Semi-finished Products for Hydroforming. steel research international 76, No. 2/3, Stahleisen GmbH, Düsseldorf. 2005.

[19] Normenausschuss Materialprüfung (NMP), Materials Testing Standards Committee. Prüfung von Klebverbindungen- Probenherstellung Berlin: Beuth Verlag GmbH; 2006.

[20] Normenausschuss Materialprüfung (NMP) MTSC. Strukturklebstoffe - Bestimmung des Scherverhaltens struktureller Klebungen - Teil 2: Scherprüfung für dicke Fügeteile (ISO 11003-2:2001, modifiziert); Deutsche Fassung EN 14869-2:201 Berlin: Beuth Verlag GmbH; 2011.

[21] Nutzmann M. Umformung von Mehrschichtverbundblechen für Leichtbauteile im Fahrzeugbau Aachen: Shaker Verlag; 2008.

[22] Alfano G, Crisfield MA. Finite element interface models for the delamination analysis of laminated composites: mechanical and computational issues. International Journal for Numerical Methods in Engineering. 2001.

[23] Hashimoto K, Ohwue T, Takita M. Formability of steel-plastic laminated sheets. In Group 1bcotIDDR. Controlling sheet metal forming processes. Metals Park, Ohio: ASM International; 1988.

[24] Takiguchi M, Yoshida F. Effect of Forming Speed on Plastic Bending of Adhesively Bonded Sheet Metals. JSME International Journal Series A. 2004.

[25] Engel B, Buhl J. Roll Forming of Vibration-Damping Composite Sheets. In The 8th international Conference and Workshop on numerical simulation of 3d sheet metal forming processes.: AIP Conference Proceedings, Volume 1383, pp. 733-741; 2011.

[26] Banks J. AutoSimulations, Inc., Atlanta, GA 30067, U.S.A. Introduction to simulation. In P. A. Farrington HBNDTSaGWEe. Proceedings of the 1999 Winter Simulation Conference. Atlanta,: Print ISBN: 0-7803-5780-9; 1999.

[27] M. Weiss, B. F. Rolfe, M. Dingle, J. L. Duncan. Elastic Bending of Steel-Polymer-Steel (SPS) Laminates to a Constant Curvature. Journal of Applied MechANICS. 2006.

[28] Takiguchi M, Yoshida F. Deformation Characteristics and Delamination Strength of Adhesively Bonded Aluminium Alloy Sheet under Plastic Bending. JSME International Journal. 2003.

[29] Hudayari Ġ. Theoretische und experimentelle Untersuchungen zum Walzrunden von plattierten und nicht-plattierten Grobblechen Siegen : Höpner und Göttert; 2001.

[30] Fritzen CP. Technische Mechanik II (Elastomechanik) Siegen: UNI SIEGEN; 2006.

[31] Kreulitsch, H. Voest-Alpine Stahl Linz GmbH. Formgebung von Blechen und Bändern durch Biegen Wien: Springer; 1995.

[32] Lerch R, Sessler G, Wolf D. Technische Akustik: Grundlagen Und Anwendungen Heidelberg: Springer; 2008.

[33] Möser M, Heckl M, Kropp W, Cremer L. Körperschall: Physikalische Grundlagen und Technische Anwendung Heidelberg: Springer; 2010.

[34] Pflüger M, Brandl F, Bernhard U, Feitzelmayer K. Fahrzeugakustik Wien New York: Springer; 2010.

[35] Knaebel M, Jäger H, Mastel R. Technische Schwingungslehre Wiesbaden: Vieweg+Teubner I GWV Fachverlage GmbH; 2009.

[36] Ross D, Ungar E, Kerwin E.. Damping of Flexural Vibrations by Means of Viscoelastic Laminate. Structural Damping. 1959.

[37] Mead DJ, Markus S. The forced vibration of a three-layer, damped sandwich beam with arbitrary boundary conditions. Journal of Sound and Vibration. 1969.

[38] Sadasiva Rao YVK, Nakra BC. Vibrations of Unsymmetrical Sandwich Beams and Plates With Viscoelastic Cores. Journal Sound Vibrations. 1974.

[39] Mead DJ. A comparison of some equations for the flexural vibration of damped sandwich beams. Journal Sound Vibration. 1982.

[40] Dewangan P. Passive viscoelastic conatrained layer damping for structural application. National Institude of Technology Rourkela. 2009.

Tools

Towards Benign Metal-Forming: The Assessment of the Environmental Performance of Metal-Sheet Forming Processes

Marta Oliveira

Additional information is available at the end of the chapter

1. Introduction

In the last decade, significant attention has been devoted to the assessment of the environmental impact of manufacturing processes and machine-tools, defining the most important factors to be covered and proposing methodologies to support the analysis of their individual contributions. The work published has established that the environmental impact of a manufacturing process is mainly affected by the consumption of 3 types of resources, namely:

1. The full set of resources used to obtain the machine-tool, accounted as input-output substances associated to the components production and their assembly (materials- and manufacturing processes-related);
2. The electricity required during operation, accounted as the specific process energy (SPE) related to the main functionality of the machine-tool;
3. Other process- or operation-related resources, apart from electricity, accounted as input-output substances associated to the use of the machine-tool (consumed directly in the process, by auxiliary systems during operation or in maintenance operations).

Most of the studies dealing with this triangular perspective focus the analysis of chipping processes (Dietmair & Verl, 2010; EBM, 2010; Gutowski et al., 2006; Kuhrke et al., 2010; Pusavec et al., 2010a,b; Rajemi et al., 2010). Other pure metal forming processes, such as chipless-shaping processes, typically involve no significant material waste or consumables usage, and the savings on the electrical consumption of the machine-tool become the dominant factor to analyse during the use-stage (Santos et al, 2011). As advanced by Gutowski (Gutowski et al., 2006), the total energy required for operation of a machine-tool is not constant, as many life-cycle assessment (LCA) tools assume, and instead the system total electricity consumption, $P_{active,System}$, should be decomposed in a fixed and a variable parts, according to Eq. (1):

$$P_{active,System} = P_0 + k\dot{v} \qquad (1)$$

where P_0 is the fixed part corresponding to the total stand-by power [kW], $\dot{v}$ is the rate of material processing, typically in cm³/s, and k is a constant provided in kJ/cm³.

Additionally, from Eq. (1), the SPE would be built as indicated in Eq. (2):

$$B_{elect} = \frac{P_0}{\dot{v}} + k \qquad (2)$$

While the constant part is used to insure the active response of sub-systems, such as driving controls, exhaustion or cooling apparatus, and is independent of whether or not a part is being produced, the variable part corresponds to the energy needed to produce a work-piece and is typically proportional to the amount of material being processed or to the type of work.

As demonstrated by Santos (Santos et al, 2011) for pure forming processes with discrete loading, such as bending, the maximum value of the variable part is limited by the machine characteristics, affecting the throughput. On the other hand, this is the theoretically constant value to which the SPE model would tend to, considering the fixed consumption would be shared by an infinite throughput. In real scenarios, the rate between constant and variable contributions associated to a production cycle, as well as their respective values, is mostly dependent on the system technology. However, the full implications of technology to the environmental profile of the machine should be attained in terms of the contributing triangle referred and not only the energy-consumption during use of the machine-tool.

Another point is the guiding for environmental improvement, as the environmental impact assessment requires the application of specific methods and tools. Life-cycle assessment (LCA) is the reference tool for environmental profiling of products and processes, as it is the most effective tool for this purpose and permits the most advanced environmental analysis possible. Every LCA methods use qualitative, quantitative or semi-quantitative analysis, although the quantitative form is considered more suitable for detailed LCA studies (Curran, 1996; Hochschorner, 2003). However, LCA tools can be time and work consuming and thus have significant costs.

In recent years, there has been a trend for the development of simplified methods for LCA. These are quantitative or semi-quantitative methodologies aiming to give quick answers and suggestions. Although these methods tend to be very universal and wide-ranging, given the broad applicability of these methodologies and the strong emergence of its use, they have a strong customization potential. In fact, these simplification techniques can be adapted to provide 'customized' or 'tailor-made' perspectives in studies of specific systems or sectors, enabling to include system-specific principles and practices more relevant and appropriate to the interested LCA end-user, while still producing valid and robust results, and keeping the LCA basic conditions regarding scope and methodology (Bala, 2010; Curran, 1996). In line with this, Hochschorner (Hochschorner et al, 2003) highlighted the importance of the method applicability to the field of application as the most important selection criteria of the proper LCA method to adopt, in order to deliver the required information.

Regarding this, it is important to highlight the interesting work followed by the CO2PE initiative (Kellens et al., 2012), which has been working on the definition of a methodology for systematic analysis and improvement of manufacturing unit process life-cycle inventory (UPLCI), i.e. on the deep analysis and quantification of the manufacturing processes environmental impact. To ensure optimal reproducibility and applicability, documentation guidelines for data and metadata are included in this approach. Guidance on the definition of a functional unit and a reference flow as well as on the determination of system boundaries meets the generic LCA goal and scope definition requirements of ISO 14040 and ISO 14044. Developed with the purpose to provide high-quality life-cycle inventory (LCI) data for manufacturing unit processes, this work seems to fit the needs of methodology standardization for the machine-tool use-stage analysis.

This chapter provides an overview on the assessment of the environmental impact of metal-forming processes and machine-tools. The most important factors to be considered are discussed and some methodologies supporting the analysis of the individual contribution of energy consumption during process are presented, as this is still considered as the main detractor. Process categorization criteria and accurate modelling of the energy consumption per category are here highlighted as the basis for high quality quantitative inventory data to achieve reliable environmental profiling. Pure metal forming processes, such as bending, are covered. Overall and sub-systems accounting strategies are presented, using the case of Laser cutting as an example of multiple sub-systems with similar contribution to the total energy consumption. The main findings and conclusions, as well as some strategies favouring the environmental performance of the manufacturing processes, are here discussed.

2. Methodologies used

The strategies for improvement expected from the environmental profile assessment of a process shall be based on the comparative analysis between alternative production scenarios. Relevant technologies and application ranges have been selected, considering the respective technology/application market share. Comparative studies were followed with a pre-defined job and application ranges (material, shape, process quality,…), and under similar utilization modes.

Regarding data collection, particularly on preliminary process evaluations, it seemed realistic to start focusing on energy consumption and any main consumable. The same measuring system and accounting methods were used throughout the comparative studies followed.

Special attention was dedicated to the accounting methods and assessment methodology to use. A wide discussion has been followed about the limitations of the non-standardized methodologies and the impact of the quality of the inventory and main indicators to the reliability and standardization of environmental profile assessment results. As presented in a previous work (Azevedo et al, 2010), for the purpose of the analysis of the environmental profile of a machine-tool, the contribution of the different main inputs to the overall impact shall be analysed relatively to each other. On the other hand, the detailed analysis per environmental impact category should consider absolute values, in order to reveal those categories to which the process is potentially more detrimental to, and which main input contributes most to it, in order to properly inform about the real extent of the impact. This was also the strategy here adopted.

2.1. Data collection

Manufacturing processes analysed were metal bending and Laser-cutting. In both cases, no other process- or operation-related resources, apart from electricity, were consumed, but the influence of the hydraulic oil needs during the equipment's lifetime was considered in the LCA analysis of a conventional press-brake used for bending. In the case of the Laser cutting machine, with individualised sub-systems, power consumption measurements were followed in parallel for the 3 main sub-systems, namely the Laser source, the chiller and the control unit (Oliveira et al, 2010).

The energy consumption data were acquired with a Janitza Power Analyzer, model UMG 604, a measuring system able to measure and calculate multiple electrical variables on 3-phase AC systems. The system was configured to record current, voltage, and power factor per line every 1 s, installed in the machine-tool's electrical cabinet; accounts over 24 h for each test have been followed. In the used configuration, the system is capable to measure low-voltage systems up to 300 V conductor to earth and currents up to 60 A, with maximum measuring uncertainties of ±0.50 V and ±0.15 A, respectively, over long periods of time (Santos et al, 2011).

The yearly consumption of hydraulic oil was taken from the Preventive Maintenance Plan provided by the machine-tools local manufacturer involved in the time-studies (Adira S.A., 2010).

2.2. Accounting and LCA methodologies

As advanced, there are still no specific tools and methodologies for the characterization of the environmental profile of a manufacturing process. This is being taken by research groups and associations, such as the CO2PE initiative previously referred, but still much as to be done on the standardization of methods and quality of the inventory data currently available. As referred, the particular system technology being used for the process, i.e., the type of machine-tool available, as well as the utilization mode during production, also plays an important role on the accounting of energy consumption during process, although this has been neglected on the common LCI databases.

All environmental analysis generated in this framework was followed by application of the Eco-Indicator 99 (H,A) method, using SimaPro 7.0 with Ecoinvent 2.0 unit processes as LCI database (Pré-consultants, 2010a,b). LCI datasets of secondary metals have been used whenever applicable on the equipment-related resources. The main inputs related to the use phase, electricity and hydraulic oil, were distinguished as different use phases, to assist their individual impact during the analysis. The LCA outcome results from Single Score analysis.

3. Critical factors in the analysis of main environmental impact contributors

Regarding the SPE contribution, the differentiation between metal manufacturing processes involving material removal and deposition from those pure forming operations, understood as discrete loading operations, has been proposed (Santos et al., 2011). In the different studies supporting this work, comparison and modelling of the electricity consumption data during process with systems of different technologies, and the influence of production use scenarios, were discussed based on time studies followed at industrial users. For discrete production cycle operations, such as bending, the definition of a specific exergy reference unit was proposed, since the units typically associated to manufacturing processes, generally described per unit of material processed, were considered not suitable. In this work, direct process categorization criteria such as system technology, maximum loading capacity and production scenario have been proposed.

On the other hand, in what refers to system technology, overall vs sub-systems (energy-consuming or not) strategies for data collection and accounting were adopted. For bending, the overall approach was used in the analysis of the pess-brakes, while for the laser cutter, parallel analysis of 3 main sub-systems was followed. This later case is in line with the current trend to more efficient power technologies and modular design, with no single dominating consumer sub-system but on a set of sub-systems with comparable energy consumption levels, which justifies the sub-system approach. In what concerns the SPE assessment, this is definitely the approach to adopt targeting the identification of main contributors, even if the total SPE value is the one to be final accounted. This problematic is patent on the following case-studies analysed.

3.1. System technology impact: Sheet metal bending

When compared to other manufacturing processes, such as chipping, coating or cutting, the effective loading time per production cycle during bending is relatively short and the specific rate at which the load is applied is not a significant process parameter. In turn, the analysis of the energy needs should focus on effective energy values in alternative to the time-dependent power parameter value.

The scans in Fig. 1 expose the referred discrete loading character inherent to bending, and the influence of machine-tool technology in the temporal evolution of power and energy consumed per operation cycle. In the case of bending, and according to the time scans presented in Fig. 1, the total energy consumption per bending cycle can be modelled according Eq. (3):

$$B_{elect} = P_{idle} \cdot \Delta t_{idle} + B_{approach} + P_{bending} \cdot \Delta t_{bending} + B_{return} \tag{3}$$

where, P_{idle} is the active power consumed during stand-by mode, being technology-related, and $P_{bending}$ is the active power consumed during loading, here proposed to be modelled according to Eq. (4):

$$P_{bending} = P_{0-bending} + c \cdot F_{bending} \tag{4}$$

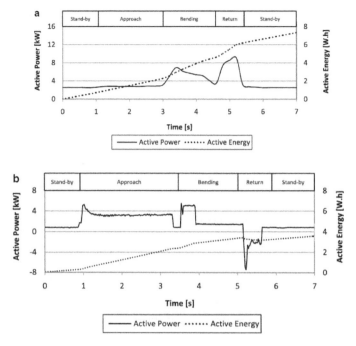

Figure 1. Examples of active power and active energy per bending cycle recorded for: a) a hydraulic press-brake and b) an all-electric press-brake.

This model is built according to Eq. (1) proposed by Gutowski (Gutowski et al., 2006), but adapted for discrete loading manufacturing processes, where, $P_{0\_bending}$ stands for the power required at a zero load cycle, c is a constant in kW/t, corresponding to the positive power rate needed to sustain a theoretical continuous load increase, and $F_{bending}$ would take the real load value pre-defined for production.

However, for such discrete loading operations, the process rate must be described as a function of the frequency of production cycles, instead of the amount of material removed or being processed. In this perspective, Eq. (2) should take the form proposed in Eq. (5):

$$B_{elect} = \frac{P_{idle}}{n} + q \tag{5}$$

where n corresponds to the throughput in cycles/h.

Following the previous discussion on the fixed and variable contributions of the specific process energy, constant q represents the cycle peak energy obtained at a pre-defined process load and loading time, excluding the fixed contribution, while the variable contribution $\frac{P_{idle}}{n}$ considers the total cycle time, which is dependent on the production throughput, in opposition to the loading time in the chipping processes. From this, it can be concluded that the actual energy and usage of each machine are essential to estimate the SPE required for bending operation or any other discrete loading operations.

The SPE as a function of the throughput was estimated for a set of hydraulic and all-eletric press-brakes based on real consumption data measured directly on the machines. Figure 2 shows the estimation models obtained for all machines, working at the highest used loading capacity during the study and with a maximum throughput value of 720 cycles/h, as this is the theoretical limit for a machine working continuously at the smallest cycle time observed (5 s).

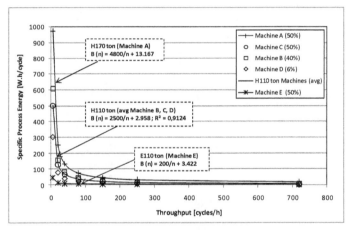

Figure 2. Specific process energy (SPE) during bending as a function of throughput, obtained from energy consumption data measured directly on a set of bending machines (n: throughput [cycles/h]; Hxxx: hydraulic technology and Exxx: all-electric technology, indicating the respective maximum bending capacity; xx%: capacity loading tested in a specific machine).

In this perspective, the following principles were proposed for the modelling of the SPE during metal-sheet bending:

- The reference unit to use is the production cycle;
- The main driving technology must be used as process categorization criteria;
- Parameterization of the SPE, for each category, is a function of the throughput.

From this, the categorization criteria proposed to be used for bending and similar discrete chipless-shaping manufacturing operations in general are the type of operation, technology, maximum loading capacity and usage scenario. Regarding the usage, typical throughput values for 3 main usage scenarios installed for bending (robot-assisted, manual-intensive and manual-discrete) have been appointed. Table 1 resumes the energy consumption values per bending technology category and usage scenario proposed to be considered for the estimation of electricity consumption related to the bending operation. These values can be used in LCI databases, in alternative to the theoretical values typically adopted, often even associated to generic manufacturing work (Ecoinvent Centre, 2010). These data should be applicable to all types of material to be worked, as they are categorized on a bending capacity basis. The potential energy savings related to the selection of the driving system technology and the motor rated power installed were quantified from the estimated SPE values and are also here included.

Specific Process Energy (SPE) [W.h/cycle]	Equipment Usage Scenarios		
Technology Category	Manual-discrete (n=20)	Manual-intensive (n=80)	Robot-assisted (n=250)
I. Bending, hydraulic, 170 t	253.2	73.2	32.4
II. Bending, hydraulic, 110 t	128.0	34.2	13.0
III. Bending, electric, 100 t	13.4	5.9	4.2
Potential energy savings			
Technology-related (III vs II)	90%	83%	67%
Motor rated power-related (II vs I)	49%	53%	60%

Table 1. Specific process energy values related to bending operation, as a function of technology and usage scenario, and comparative analysis in the form of potential energy savings.

Particularly for irregular and/or low usage scenarios, the electric-based drive technology is to be recommended, as this might lead to energy savings of about 90% when compared to an all-hydraulic system, for a similar loading capacity machine, while the potential savings tend to be reduced for more intense usage scenarios. Nevertheless, even for the highest robot-assisted production scenarios, energy savings as high as 67% could be achieved with electrically-driven systems when compared to the hydraulic ones, for similar loading capacities installed.

As advanced, apart from the SPE, technology also determines the type and amount of other consumables during operation. In the case of a hydraulic press-brake, hydraulic oil is a technology-specific resource essential for its operation. The environmental impact profile related to the oil consumption is significantly affected by its no-renewable character, as this is a standard crude oil by-product, typically incinerated at the end-of-life. Figure 3 shows the contributions of the Assembly-phase and Use-phase inputs (Electricity and Oil) to the environmental profile of a hydraulic press-brake, described per different middle-point impact categories. A lifetime of about 15 years was assumed, and the contribution of an end-of-life scenario has here not been accessed. The most probable machine-tool end-of-life scenario is reutilization as second-hand which, in practice, would represent an extension of the lifetime, reflected by an increase on the use-phase inputs, i.e. SPE and hydraulic oil contributions.

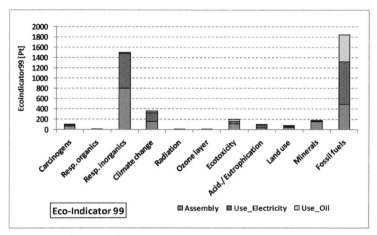

Figure 3. LCA results of the Assembly-phase and Use-phase (Electricity and Oil) inputs used for an hydraulic press-brake, per impact category (middle-point).

These results make evident that the contribution of the hydraulic oil used to the environmental impact of such machines should not be neglected. The relative contribution of the oil consumption to the global environmental impact of the machine is higher than 20%, in at least 2 of the 11 middle-point impact categories analysed. The main impact of the oil consumption, in absolute value of the indicator, is on the depletion of fossil fuels, and, in this category, the impact is similar to that of the total of assembly resources incorporated in the machine, which is considered quite significant.

3.2. Sub-system approach on SPE accounting: Sheet metal cutting

In the case of a Laser-cutter, the Laser energy source and the cooling system determine its overall integration of components and energy consumption and, thus, their analysis is essential to characterize the environmental profile of the Laser cutting process. The recent Fiber Laser (FL) technology has been compared with the well-established CO_2-Laser (CO2). While the technical benefits related to the integration of fewer components are intuitive to promote the environmental performance of the FL technology, the strategies for energy savings are not that evident and there is still some room for improvement (Oliveira et al, 2011). Besides the Laser and cooling units, other important electrically fed subsystems, such as the general control unit (including the motorized head positioning) and the exhaustion system, should not be neglected on the analysis of the energy demand of a Laser cutter, although this later has not been actually measured in this work.

Figure 4 presents the SPE results obtained from analysing 3 equipments of FL and CO2 technologies, in similar utilization conditions and for a similar job (1 mm steel sheet). The technology influence on the SPE is also here made evident.

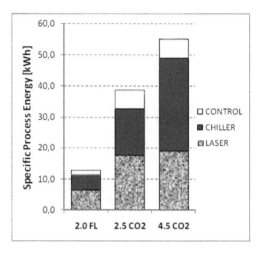

Figure 4. Specific process energy (SPE) observed for the Laser cutting process, showing relative contribution of the main sub-systems per equipment type (FL: Fiber Laser technology; CO2: CO2-Laser technology). Reference is 1 mm carbon steel sheet processing for 1 h.

Regarding the individual contribution of each main sub-systems in CO2 equipment, the chiller unit of a 4.5 kW machine was seen to be responsible for more than 50% of the energy demand, contradicting the assumption that the Laser source dominates the total energy consumption of these machines (Devoldere et al, 2006). In what concerns to the FL equipment, the energy benefits were clear:

- The higher global efficiency of the FL technology provided for a reduction of at least 1/2 on the SPE of the cutting process of thin sheets;
- The energy consumptions of both Laser and chiller units were significantly reduced. Lower cooling requirements result from the lower Laser input power and reduced energy losses.
- The contributions of the 2 main electrical sub-systems (Laser and chiller) are expected to bein the same order to that of the control unit.

In resume, FL technology brings down the Laser and chiller energy needs, to the consumption level of standard control/motion units, or even exhaustion systems. In such a scenario of no predominant contributor, and targeting to maximize energy efficiency, all sub-systems apart from the Laser unit must be accurately specified.

4. Best practices and improvement opportunities

Although energy efficiency improvements adopted along the last 20 years were seen to reduce energy requirements of machine-tools in approximately 50%, the basic guidelines for energy savings during process, such as the specification of most energy efficient components and guidelines for effective energy management during machine-tool processing are still not established. The examples given in the previous section support the two strategies generally proposed to improve energy efficiency: (1) the conversion of hydraulic to all electric systems and (2) the maximization of the rate at which the physical mechanism can perform the desired operation, i.e., the optimization of machine usage.

It must be noted that awareness of the manufacturing end-user regarding the importance of energy management should be strongly enhanced. Although widely discussed in different areas, this is a topic that the manufacturing user tends to neglect, regarding each individual machine on his plant, particularly in what concerns technology, process and usage strategies. Independently on the many possible solutions targeting the automatic control of the machine-tool, the user's perception surely determines this optimization. Enabling the user to obtain detailed and real-time data about the energy consumption of the manufacturing process is essential to accomplish the optimization of the machine-tool environmental profile during the use stage, as the user must be actively involved in this process. It is on the side of the machine-tool manufacturer to preview and implement this. On the other hand, and apart from all criteria behind the selection of each individual sub-system on the machine, including its technology, it is on the manufacturer side to match the power demand profile of the main energy-consuming sub-systems integrated, in what concerns the power consumption of the sub-systems, as realised from the Laser cutting study followed.

However, as referred at start, the technology-related improvement potential of a manufacturing process towards benign metal forming must be sustained by an integrated perspective, mainly presenting energy-related technical solutions but not-only, as the contribution of the 3 types of resources listed above is affected, namely the assembly resources, the energy consumption during use and the other consumables related to machine operation, such as the influence of the hydraulic oil to the environmental impact of the bending process, as here demonstrated. Combining these perspectives, and in view of the discussed influence of the machine-tool technology on the environmental impact of the manufacturing process, special attention is here given to the assembled sub-systems of the machine-tool, and particularly to the materials incorporated on these, in which steel has traditionally been dominating. On the other hand, the change in steel pricing policy and current increasing steel cost are pressing overheads and margins at the machine-tools manufacturers and their components suppliers. As the need for alternative materials, less subjected to such market variations, becomes more evident, technical targets, process quality and environmental profile might be compromised. In addition, market has been specifically requiring performance increase, in the sense of higher stiffness, dimensional stability, ease of manufacturing, good dampening properties and high mass to avoid rigid body movements. Some examples of high potential actions enhancing benign metal forming currently being developed and adopted are pointed out in the next sub-sections.

4.1. Detailed analysis of assembled sub-systems components

Regarding the assembled sub-systems of the machine-tool, and considering the trend for all electric or electromagnetic versions of these, particular attention should be given to the use of advanced functional materials, particularly composites, and the increased use of additional electronic components, as these typically includes higher amounts of hazardous materials or raw materials which are hard to recover. Also on this analysis, the sub-system approach for improvement is recommended in order to favour a finer analysis of all components. In fact, while the significant impact of a housing material can be more evident from the volumetric contribution of the component, only a detailed sub-system analysis can insure that the determinant impact of a small volume component based on a hazardous material would not be missed. Although some mandatory related legislation is established for electronic components, the amount and combination of substances in a multi-component electromechanical sub-system is still relying on the environmentally conscious of the sub-system manufacturer.

4.2. Mass reduction of moving parts

In moving sub-systems/parts of machine-tools, the current replacement of standard materials by lightweight alternative materials simultaneously reflects the trend to optimized material consumption, general material reduction and the introduction of high-performance materials, such as reinforced polymer-based composites or low-density metals. Although this trend is often pointed out as a positive factor pushing for new dynamics to the sector, the issue of the environmental cost of the introduction of these alternative materials should be carefully analysed, particularly regarding the lifetime and end-of-life disposition of such components/materials, although a lot of work is on-going regarding innovative end-of-life strategies for these materials.

4.3. Mass reduction of structural parts

In what concerns the assembly resources used for the machine-tool construction, the materials and process inputs associated with the base structure of the equipment tends to determine its environmental impact, due to its dominant volume and weight. When looking for high performance materials for high-accuracy processes/systems, innovative polymer concrete solutions, also referred as mineral casting, are being introduced to replace the typical steel welded main structure towards a performance upgrade even in the most conventional machine-tools. Technically, this solution is indicated to overcome the static and dynamic stiffness and vibration damping requirements (Erbe at al., 2008), reflected on Figure 5, but, indirectly, this has significant environmental, technical and cost benefits.

Polymer concrete compositions mainly integrate a set of mineral granulates dispersed in a polymer resin. Granulates are abundant and several companies even supply these mineral products with certified composition, granulometry and general quality specifications, which have the advantage of being market proven, reducing risk and time-to-market. Unfortunately, although they are about 90% of the total weight of the composition, these components

represent less than 20% of the cost. In fact, polymer resins are the cost-drivers in these
compositions. In general, epoxy resins are about four times more expensive than alternative
polyester resins. Depending on the quality requirements, polyester could be a preferable
choice, but they are less stable and present a higher shrinkage rate, which might be a problem
in thicker bodies, as high shrinkage might result on significant internal stressing and
subsequent cracking. Mineral casted structures can be produced in a single-step, and process
time is mainly affected by the curing process, which depends mostly on the polymer
characteristics (some products are presenting curing times up to 24 hrs at room temperature).

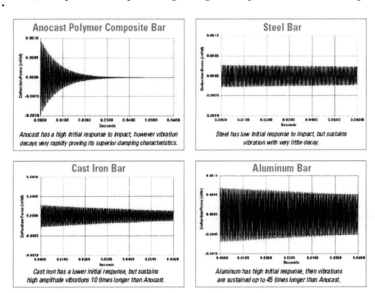

Figure 5. Comparative analysis of vibration dampening between conventional metals used in machine-
tools and a commercially available polymer composite (based on Anocast product, a Rockwell
Automation product (Rockwell, 2012)).

In turn, regarding the environmental benefits related to the introduction of polymer
concrete-based solutions, the following aspects are to be highlighted:

- Reduction of process lead-time: Mostly depending on mixing, casting and curing
  processes, mineral cast structures are substantially faster available than the traditional
  casting or welded steel parts;
- Reduction of energy consumption during assembly phase of the machine-tool: The
  energy requirements to produce a polymer concrete structure is foreseen to be about
  25% of that needed to produce an equivalent welded steel structure;
- Lifetime and potential for reuse: Mineral casting is chemically inert against aggressive
  materials such as oils, caustic solutions, acids and liquid-coolants. If crushed, it can be
  re-used as filler to the new mineral casting composition.

4.4. Selection of sub-systems technology and power matching

As concluded from the study on Laser cutting, although the sub-systems dimensioning and energy-consumption could be individually optimized for a specific range according to the application conditions, these are quite technology dependent. Besides, considering that one same machine-tool model typically operates in very distinct operating modes, it is important to insure that all sub-systems are properly synchronized in each operating condition, and the respective power-consumption profiles should be matched. This is expected to contribute significantly for the reduction on the power demand and improved efficiency of auxiliary main-systems against the main energy-source sub-system.

Besides the evidences coming from the laser case analysed, the impact of the motor power to the SPE values of the bending case also reflects some needs for improvement. Although the energy models presented are able to be tuned for different motor rated power levels integrated in press-brakes of different maximum bending capacity, this is indirectly associated to the maximum loading capacity of the machine. Obviously, in order to minimize power consumption, the motor rated power installed should be as low as possible, as made evident when comparing hydraulic systems (Table 1), where the lower motor rated power of the former resulted in over 49% energy savings when comparing a 110 t equipment with that of a 170 t equivalent. Moreover, in these hydraulic systems, where the stand-by consumption contributes significantly to the SPE, the motor power related energy savings are maximized with the increase on system usage.

5. Conclusion

The environmental impact of a manufacturing process is strongly determined by the basic features associated to the characteristics of the machine-tool selected to execute the process. The analysis of the machine-tool assembly is supported by the inventory of all substances used as components' materials or production consumables. In addition, the analysis of a specific manufacturing process corresponds to the collection of all substances used during utilization of the machine. In a full LCA of the machine-tool, this would correspond to the use-phase of the machine-tool life-cycle.

Machine-tool technology is the factor determining the assembly components type, amount and arrangement, and consequently the energy consumption profile of the machine, which rules the environmental performance of the manufacturing process. Modelling the energy consumption of a process firstly requires an adequate process categorization based on the technology of the main functional sub-systems. Other categorization criteria, such as the utilization mode, are also interesting, but more extensive work is needed to validate and reveal the most relevant sub-categories and associated environmental features. Attention should also be paid to the exergy reference unit used to define the specific process energy indicator. In the case of pure metal forming processes, which includes a set of chipless-shaping processes, some applying only discrete loads, typical units based on the amount of material removed are not appropriate. In such cases alternative units can be introduced, such as the energy per bending cycle proposed for bending.

In the future, the environmental impact of the manufacturing processes will be strongly affected by the trend for sub-systems modularisation, higher accuracy and versatility, as well as legislation and cost factors. The particular optimisation of the energy consumption of the machine-tools during process will require a strong awareness of machine-tool manufacturers and end-users, as the continuous improvement will not depend on a single measure. Proposed measures to be combined include solutions of alternative materials, either for small components or main structures, moving or structural parts, matching of sub-systems power profile and conditions of application, operating modes, maintenance needs and process chain shortening. Many high potential measures towards metal forming, and general manufacturing processes, are being revealed by dedicated groups, but extended work focused on the development and standardization of accounting and assessment methods customized for the purpose of evaluating the environmental profile of each manufacturing process category must be followed.

Author details

Marta Oliveira
INEGI – Instituto de Engenharia Mecânica e Gestão Industrial, Portugal

6. References

Azevedo M., Oliveira M. I., Pereira J.P. and Reis A. (2011), Comparison of two LCA Methodologies in the Machine-tools Environmental Performance Improvement Process, *Proceedings of 18th CIRP International Conference on Life Cycle Engineering*, Braunschweig-Germany, May 02-04.

Bala, A., Raugei, M., Benveniste, G., Gazulla, C., Fullana-i- Palmer P. (2010). Simplified tools for global warming potential evaluation: when 'good enough' is best, *International Journal of Life Cycle Assessment*, Vol. 15, pp. 489-498.

Curran, M. A., Young, S. (1996). Report from the EPA conference on Streamlining LCA, *International Journal of Life Cycle Assessment*, Vol1, No. 1, pp. 57-60.

Devoldere T., Dewulf W., Deprez W., Duflou J. R. (2008), Energy related environmental impact reduction opportunities in machine design: case study of a laser cutting machine, *Proceedings of the CIRP International Conference on Life Cycle Engineering*, Sydney, N. S. W., pp. 412-419.

Dietmair A., and Verl A. (2010). Energy consumption assessment and optimization in the design and use phase of machine tools, *Proceedings of the 17th CIRP International Conference on Life Cycle Engineering*, Anhui-China, May 19-21.

Ecoinvent Centre, Ecoinvent 2.0 database, October 2010, www.ecoinvent.ch.

Environmentally Benign Manufacturing (EBM) research group, Laboratory for Manufacturing and Productivity, Mechanical Engineering Department, MIT, January 2010, web.mit.edu/ebm/www/publications.htm.

Erbe T., Król J.; Theska R. (2008). Mineral casting as material for machine base-frames of precision machines. *Proceedings of 23rd Annual Meeting of the American Society for Precision Engineering and the Twelfth ICPE*, October 19-24, Portland-Oregon.

Gutowski T., Dahmus J. and Thiriez A. (2006). Electrical energy requirements for manufacturing processes, *Proceedings of 13th CIRP International Conference on Life Cycle Engineering*, Leuven-Belgium, May 31-Jun 2, 623-628.

Hochschorner, E., Finnveden, G. (2003). Evaluation of two simplified life cycle assessment methods, *International Journal of Life Cycle Assessment*, Vol. 8, No. 3, pp. 119-128.

Kellens K., Dewulf W., Overcash M., Hauschild M., Duflou J. R. (2012). Methodology for systematic analysis and improvement of manufacturing unit process life cycle inventory (UPLCI) - CO2PE! initiative (cooperative effort on process emissions in manufacturing) Part 1: Methodology description, *International Journal of Life Cycle Assessment*, Vol. 17, pp. 69–78, DOI 10.1007/s11367-011-0340-4.

Kuhrke B., Schrems S., Eisele C., and Abele E. (2010). Methodology to assess the energy consumption of cutting machine tools', *Proceedings of the 17th CIRP International Conference on Life Cycle Engineering*, Anhui-China, May 19-21.

Manufacturer preventive maintenance plan (2010), internal document of partner machine-tool manufacturer , October 2010, www.adira.pt.

Oliveira M. I., Santos J. P., Almeida F. G., Reis A., Pereira J.P., Rocha A. B. (2011). Impact of Laser-based Technologies in the Energy-Consumption of Metal Cutters: Comparison between commercially available System, *Proceedings of 14th International Conference on Sheet Metal*, Leuven-Belgium, Apr. 18-20, published on *Key Engineering Materials*, Vol. 473, pp. 809-815, Trans Tech Publications, Switzerland, DOI:10.4028/www.scientific.net/KEM.473.809.

Pré-consultants, Report: 'The Eco-indicator 99 - A damage oriented method for Life Cycle Impact Assessment', October 2011, www.pre.nl,.

Pré-consultants, Simapro 7.0 software, October 2011, www.pre.nl.

Pusavec F., Krajnik P., Kopac J. (2010). Transitioning to sustainable production – Part I: application on machining technologies. *Journal of Cleaner Production*, Vol. 18, pp. 174-184.

Pusavec F., Krajnik P., Kopac J. (2010). Transitioning to sustainable production – Part II: evaluation of sustainable machining technologies. *Journal of Cleaner Production*, Vol. 18, pp. 1211-1221.

Rajemi M. F., Mativenga P. T., Aramcharoen A. (2010). Sustainable machining: selection of optimum turning conditions based on minimum energy considerations. *Journal of Cleaner Production*, Vol. 18, pp. 1059-1065.

Rockwell Automation, Anorad General Catalog, p. 149, April 2012, http://www.rockwellautomation.com/anorad/.

Santos J. P., Oliveira M. I., Almeida F. G., Reis A., Pereira J.P. and Rocha A. B. (2011). Improving the environmental performance of machine-tools: influence of technology and throughput on the electrical energy consumption of a press-brake, *Journal of Cleaner Production*, Vol. 19, No. 4, pp. 356-364, DOI: 10.1016/j.jclepro. 2010.10.009.

Chapter 6

Impact of Surface Topography of Tools and Materials in Micro-Sheet Metal Forming

Tetsuhide Shimizu, Ming Yang and Ken-ichi Manabe

Additional information is available at the end of the chapter

1. Introduction

Microforming technology has been receiving much attention as one of the most economical mass production methods for micro components (Geiger et al., 2001). Especially, metal foils have the great advantage to produce high-aspect three-dimensional shapes by miniaturizing the process dimensions of sheet metal forming technologies.

Since the relative ratio of the surface area to the volume of metal foils becomes significantly larger with miniaturization, tribological behaviour is of great significance for the micro-sheet formability. Over the last decade, basic researches of the size effect of tribology in microforming have been performed worldwide (Vollertsen et al., 2009). One of the representative reports of scale effect in bulk metal forming is the double-cup-extrusion (DCE) test (Engel, 2006). By scaling the diameter of CuZn15 specimen from 4mm to 0.5mm, the scaled DCE test was conducted. Identified with a FE analysis, the significant increase of the friction factor m with decreasing the scale was confirmed. Another reports regard to the size effect of coefficient of friction in bulk metal forming was done by Putten et al. (Putten et al., 2007). The scaled plane strain compression tests were conducted with the aim of application on flat rolling. As similar tendency as DCE test, the friction coefficient increased with decreasing scale dimension.

The work focused on the sheet metal forming has been done by Vollertsen and Hu (Vollertsen & Hu, 2006, 2007). The strip drawing method allowing the determination of friction parameters for micro-deep drawing was developed. It was shown that the friction coefficient again increased with decreasing the size. Additionally, the scale dependent friction coefficient was determined by strip drawing test and the calculated value was introduced to the FEM simulation. Especially if the non-uniform pressure distribution was taken into account in the deep drawing simulation, relative good results were derived for different dimensions (Vollertsen et al., 2008).

These tendencies of the increasing friction coefficient with decreasing dimension in many experiment of forming process have been mainly explained by the "lubricant pocket model" (Engel, 2006). The lubricant pocket model is the only one available model for the description of size effects in lubricated friction (Vollertsen et al., 2009). The basic feature of describing the size effect based on this model is that the increasing relative ratio of OLPs (Open lubricant pockets), which cannot keep the lubricant. With decreasing the scale dimension, the relative ratio of OLPs increases and it results in the increase of friction resistance (Engel, 2006). As overviewed above, the overall investigation suggests the low effect of lubricant in micro-scale region.

From the other point of view of:

1. Dirt handling of the tiny work pieces,
2. Contamination of fine products,
3. Lubricant clogging between the micro scale clearance, and
4. Unstable formability due to the high sensitivity of the variation of lubricant quantity or the effects of meniscus and viscous forces,

the microforming process would be preferred not to use a lubricant (Aizawa et al., 2010).

In response to these findings, the activity of the application of the coating treatment on the die substrate, targeting the microforming, is gradually increasing. Hanada et al. fabricated a micro-die utilizing chemical vapour deposition (CVD) diamond coating (Hanada et al, 2003). The surface roughness of the diamond dies was approximately 10 nm, and the diamond dies showed good lubricating ability in the microcoining of polymethylmethacrylate (PMMA). Fuentes et al. investigated the tribological properties of Al thin foils (0.2mm nominal thickness) in sliding contact with PVD-coated carbide forming-tools for microforming (Fuentes et al., 2006). Uncoated, CrN-coated and WC-C-coated tools were tested, using a pin-on-disk configuration. They have shown that the sticking of Al was retarded using low friction magnetron sputtered WC-C-coated carbides. Fujimoto et al. proposed a novel surface treatment process for micro-dies (Fujimoto et al., 2006). They developed a high-energy ion beam irradiation for finishing die-surfaces and a CVD diamond-like carbon (DLC) was coated on the die surface after finished. They have succeeded to reduce the surface roughness by ion beam irradiation process and the high wear resistance of the DLC coating was demonstrated with the 50,000 shots of the microbending tests. In the recent work on the coating treatment for micro die, an impressive coating technology was proposed by Aizawa et al. (Aizawa et al., 2010). The nano-laminated DLC coating was invented and applied to improve the coated tool life, where delamination of coated layers was significantly retarded or saved by optimum interlayer and nano-scopic lamination of DLC sublayers. Micro stamping system, which included the severe wearing condition of ironing and bending step, was employed in the 10,000 shots progressive dry micro-stamping, where SKD-11 punches underwent severe wear in stamping of AISI-304 stainless sheets or pure titanium sheets.

Although these applied researches on the field of dry-microforming are well developed for the practical application, the basic research on the dry friction behaviour during microforming was quite few and it was only worked for bulk metal forming (Vollertsen et al., 2009). Krishnan et al. investigated the scale effect on dry friction by scaled forward extrusion test of brass (Cu:Zn 70:30) with the scale range of 0.57-1.33mm in outer diameter of extruded pin (Mori et al., 2007; Krishnan et al., 2007). In addition, the other author conducted the scaled ring compression test of CuZn15 alloy with different grain size. The test was scaled in the range of 0.5 to 4.25mm of ring inner diameter (Vollertsen et al., 2009). Although the purpose of those studies is to determine the size effect under dry friction, the statement for the size effect of dry friction seems to be very doubtful and there are no general explanations.

According to the tribology theory for dry contact friction (Bhushan, 2003), friction resistance is contributed by:

- Adhesion of the sliding surface,
- Deformation of surface asperities in contact,
- Plowing by wear particles and hard asperities, and
- So called, ratchet mechanism, which is due to a lateral force required for the contact asperities to climb against each other.

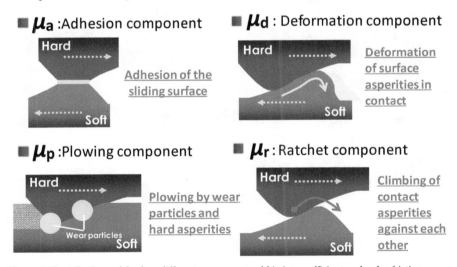

Figure 1. Contributions of the four different components of friction coefficient under dry friction

As illustrated in Fig.1, the geometry of surface asperities would contribute to the real area of contact and sliding resistance, and it is a dominant factor over the dry friction behaviour.

However, basic research on the meso-scale tribological behaviour of dry friction, such as the contact interaction of the surface asperities, is not well discussed, especially for micro-sheet metal forming (Vollertsen et al. 2009). In view of the surface functionalization and the structural optimisation, construction of the surface design guide based on the characterization of tribological behaviour under dry friction is a pressing need for the further high precision forming process design.

In view of the significant importance of tribology in micro-sheet metal forming, this chapter creates an overview of the effect of surface topography of tools and workpieces on micro-sheet formability. Starting with an introduction of the newly developed micro-deep drawing experimental system and that of a finite element simulation model with surface asperities, the impact of surface topography on the tiny micro-scale forming behaviour is discussed.

2. Development of micro-deep drawing experimental system

The experimental system for micro-deep drawing is newly developed with a goal for highly accurate forming operation with a good reproducibility in experimental results. The system consists of (a) micro-drawing die assembly, (b) desktop size miniature press machine, (c) control panel and (d) compact feeding device. Fig. 2 shows the appearance of developed micro-deep drawing experimental system.

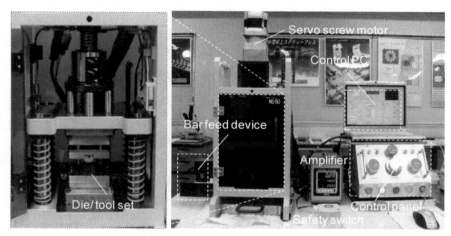

Figure 2. Appearance of micro deep drawing experimental system

2.1. System specification and design of die assembly

Developed micro-drawing die assembly is devised to improve the handling characteristics of tiny blank and to enhance the positioning accuracy of central axis of drawing process. A blanking-drawing process, which combines the blanking and drawing, is adopted and tool set is designed. The schematic illustration of micro-blanking-drawing process is shown in Fig.3. The tool of drawing die combined with the blanking punch is able to form directly from the blanking material to drawing micro-cup. In addition, blank holder force during the process is not applied for better reproducibility of the experiment. Instead of the blank holder, the constant gap between the drawing die and the blank holder is designed.

Fig.4 shows the appearance of designed micro-deep drawing die assembly. The die assembly is designed to compact palm-size with around 10cm square. Blanking punch-drawing die is mounted on the upper die, while the blanking die and the drawing punch is set up to the lower die. Furthermore, in order to measure the forming load during the process, the micro-compression load cell (TC-SR50N, TEAC Co.), which has the rating capacity of 50N, is aligned directly below the drawing punch of the lower die.

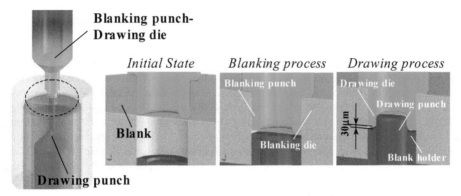

Figure 3. Schematic illustration of micro sequential blanking-drawing process

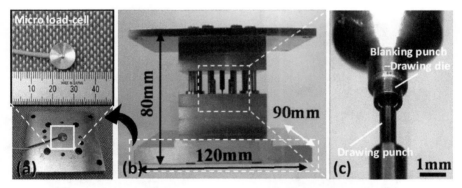

Figure 4. Appearance of designed micro drawing die assembly (a) micro compression load-cell, (b) micro drawing die assembly, (c) tool set of micro sequential blanking-drawing process

The micro-drawing die assembly is mounted on the desktop size miniature press machine (MS-50M, Seki Co.), which is custom-designed for the micro deep drawing experiment by Seki Corporation. The press machine is driven with a servomotor, which has a high motion resolution of 400nm and a maximum instantaneous velocity of 28mm/s. The motion control is based on the high precision digital displacement sensor (GT2 Keyence Co.) mounted on the press machine. In addition, a variety of press ram motion can be realized by developed controlling program, such as linear or S-curve acceleration and deceleration, inching motion at bottom dead point and motion stop at several displacement for input time length.

Moreover, the compact feeding device is newly developed for the miniature press machine. By using this device, the coiled material of metal foils can be progressively supplied during the process and the continuous transition of the forming behaviour, such as wear of tools, would be able to track in the experiment.

2.2. Fabrication of micro circular cup

By using the developed micro-deep drawing experimental system, a trial test for fabricating the micro-circular cup is carried out. As a first target of the process dimensions for the micro-deep drawing experiment, we aimed to produce the microcups with outer diameter of 500μm from the blank of 1.1 mm in diameter and of 20μm in thickness. Material used is stainless steel (JIS: SUS304-H), phosphor bronze (JIS: C5191-H), and pure titanium (JIS: TR270C-H) ultra-thin metal foils. Redrawing process is adopted to produce a cup of 700μm diameter at the first-stage and 500μm diameter at the second-stage. All microtools are made of sintered WC-Co hard alloy (JIS: V20 tungsten-carbide-cobalt alloy), machined by EDM (Electrical discharged machining) and mechanically fined polishing. By adjusting the clearance between die and punch, die corner radius and punch corner radius, the tool dimensions are determined as shown in Fig. 5.

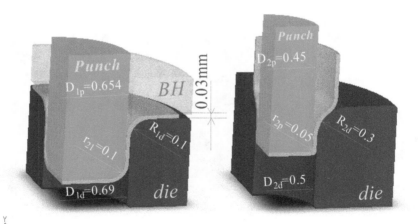

Figure 5. Schematic illustration of tool dimensions for micro-drawing experiment

Fig.6 shows the fabricated microcups on a forefinger and the SEM images of the first-stage and second-stage drawn microcups. The tiny microcups are fabricated successfully by the microforming technology. The micro-deep drawing process is realized by a high-precision blanking-drawing technique, and more than 100 microcups are produced with good reproducibility for every three kinds of materials, as shown in Fig. 7.

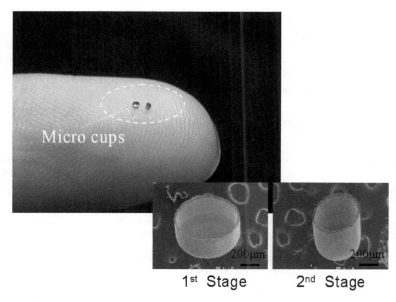

Figure 6. Appearance of drawn microcups on a forefinger and its SEM image

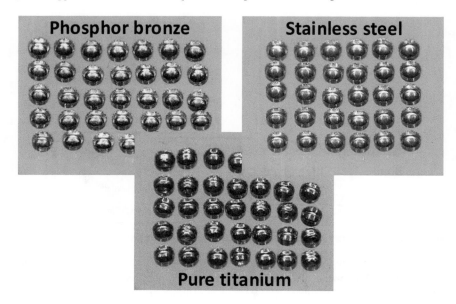

Figure 7. Appearance of drawn microcups with good reproducibility in three different materials, phosphor bronze, stainless steel, and pure titanium

3. FE analysis of surface roughness model

In order to analyse the influence of the surface asperities in wider range of roughness conditions of the tools and the materials, finite element (FE) analysis of the micro-deep drawing are carried out. To model from the microscopic roughness asperities to macroscopic material deformation behaviour, an advanced model with surface roughness is proposed.

3.1. Surface roughness model

A cyclic triangular concavo-convex configuration is applied to the FE model of micro- deep drawing. Fig.8 shows a schematic view of the surface roughness model. The blank diameter is 1.1mm and 0.02mm in thickness. The axisymmetric FE mesh sizes of quadrilateral four-node elements are 1μm×0.5μm. A cyclic surface geometry is modelled as the height of the profile, R_z, and the pitch, P, both of which are variables. This surface model is constructed on whole surface elements of blank and tools. Some of the virtues to consider the surface asperity in the FE model are as follows (Shimizu et al., 2009):

1. It is possible to input the geometry of surface asperity as a parameter of the process.
2. It is possible to represent the local deformation of surface asperity caused by contact between the tool and the material.
3. It is possible to investigate the effect of local contact behaviour on global deformation properties, such as formability and forming accuracy.

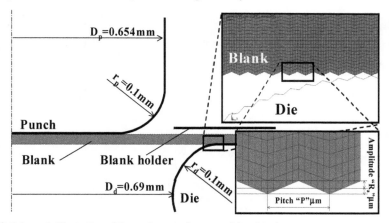

Figure 8. Schematic illustration of the surface roughness model and tool geometry

Fig.9 shows the distribution of equivalent plastic strain during the process with a surface roughness model (Shimizu et al., 2009). As shown in Fig. 9, plastic strain is observed at each local surface asperity. Particularly at a part of the sliding on the die corner and that of ironing, the amount of plastic strain of blank surface asperities is significant, due to the severe contact with the tool surface asperities.

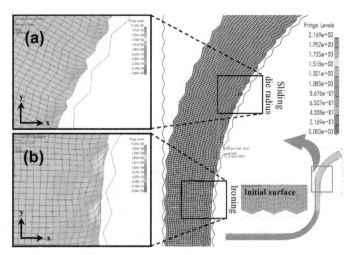

Figure 9. Distribution of equivalent plastic strain and close up images on part of (a) sliding with die corner and (b) ironing

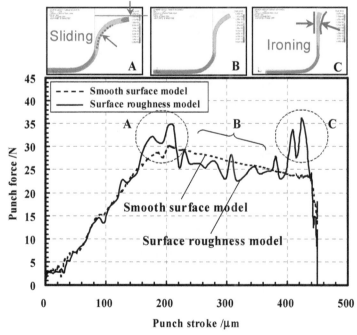

Figure 10. Comparison between punch force-stroke curves of "surface roughness model" and conventional "smooth surface model"

By comparing the surface roughness model with the smooth surface model, which is the conventional FE model with no roughness asperities, the difference with the surface roughness model can be recognized. As shown in Fig. 10, two peaks of maximum punch force are observed for the surface roughness model, while these peaks could not be observed in the smooth surface model. The first peak is under the sliding with a die corner (A) and the second peak is due to the ironing (C). Hence, by considering the local surface asperity on the model, effect of friction resistance of each surface asperity on global forming force are demonstrated.

3.2. Validation of surface roughness model

To validate the surface roughness model, the results of FE analysis are compared with the experimental results. As a representative evaluation, the micro-drawn cups of stainless steel (Fig.6) are precisely evaluated. As an evaluate item to investigate the deformation behaviour during forming process, thickness strain distribution of the drawn cup is precisely evaluated by digital image processing. Additionally, surface roughness of the drawn cups is also measured by confocal laser scanning microscope (LEXT OLS-3000, Olympus Co.).

The simulation is carried out with an explicit dynamic finite element code, LS-DYNA ver.970. Tools and blank dimensions are the same as those in the experiment (Fig.5). The blank model is assumed as isotropic elastoplastic body and it is modeled as an nth power hardening law material ($\sigma=F\cdot\varepsilon^n$; σ: flow stress, F: strength coefficient, ε: true strain, n: n value). Tools such as a punch, a die and a blank holder are assumed as rigid bodies. All models are considered axisymmetric. Table 1 shows the mechanical properties of the blank and tools used in the simulation. The static and kinetic friction coefficients between the blank and the tools are assumed to be 0.05 and 0.03, respectively.

	Blank	Tool
Mass density (g·μm⁻¹)	8.00×10^{-11}	8.00×10^{-7}
Young's modulus (GPa)	177	206
Poisson's ratio	0.3	0.3
F-value (GPa)	1.55	-
n-value	0.14	-

Table 1. Mechanical properties used in FE simulation

Fig.11 shows the comparison data of thickness strain distribution of both experiment and FE analysis (Manabe et al., 2008). General tendency of thickness variations in deep drawing process, such as the reduction at cup corner radius and the increase at cup edge, can be observed. Hence, the validation from the point of geometry and dimension of the model is demonstrated.

Fig.12 shows the comparison data of the surface roughness measured at defined 6 point of redrawn micro cups (Manabe et al., 2008). Although almost whole values are corresponded to the result of FE analysis, only the difference at the inner surface is remarkable. From the observation of inner surface roughness of drawn cup, the surface roughening phenomenon seems to largely affect the surface quality of micro drawn cup. Therefore, in order to evaluate the surface quality more precisely, the consideration of the roughening phenomenon would be required.

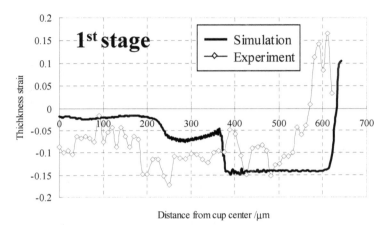

Figure 11. Comparison data of thickness strain distributions of micro-drawn cup

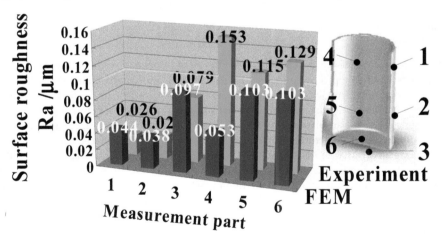

Figure 12. Comparison of surface roughness on each part of micro-drawn cup

Thus, by comparison of the experimental and FE simulation results for micro-deep drawing, the validation of surface roughness model are qualitatively demonstrated for the study on the effect of surface roughness for micro-deep drawing (Manabe et al. 2008).

4. Effect of surface topography on micro-deep drawabilty

By using the developed micro-deep drawing experimental set-up and proposed surface roughness model, the effect of the surface topography of tools and materials are experimentally and numerically investigated. The impact and sensitivity of the difference in surface properties during micro-scaled forming is demonstrated.

4.1. Effect of tool surface properties

Firstly, the effect of the tool surface properties is experimentally investigated. In order to study the sensitivity of tool surface properties on microformablity, micro-deep drawing test is carried out with three different tools with the different surface properties. By comparing the forming results of the forming force and the surface quality of the micro-drawn cups, the effect of tool surface properties are discussed.

4.1.1. Materials and experimental conditions

The material used was stainless steel ultra-thin foils (JIS: SUS304-H, 20µm in thickness). Table 2 shows the mechanical properties of the used stainless steel ultra-thin foils, which was obtained by the tensile test.

Young's modulus	Yield stress	Tensile strength	Elongation
193[GPa]	1192[MPa]	1460[MPa]	1.5[%]

Table 2. Mechanical properties of stainless steel foils (JIS: SUS304-H, 20µm in thickness) used in the experiment.

As for the micro tools for the test, the same tools as mentioned in section 2.2 were used. In order to simplify the process, only the first stage process was compared between different surface conditions. To fabricate the microtools of different surface properties, air-blasting treatment and ion beam irradiation treatment were conducted for both micro-drawing die and punch. For the air-blasting treatment, glass powder of 53-63µm size was irradiated for 5-10min on the microtool surface to produce the rough surface. An air-blasting apparatus PNEUMA BLASTER SGK-DT (FUJI Manufacturing Co.) was used for the treatment. The ion-irradiation was performed by electron cyclotron ion shower system, EIS-200ER (ELIONIX Inc.), on the condition as shown in Table 3. Fig.13 shows the 3D surface images and the surface profile data of untreated, air-blasted and ion-irradiated micro-drawing punch obtained by laser scanning confocal microscope (LEXT OLS-3000, Olympus Co.).

Ion gas	Acceleration Voltage	Vacuum	Ion current density	Irradiation time	Irradiation angle
Ar	800eV	1×10^{-4}Pa	1.2mA/cm^2	5-10min	45°

Table 3. Conditions for ion beam irradiation

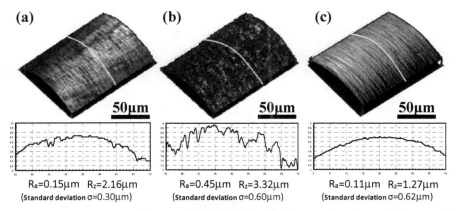

Figure 13. 3D surface images of the micro drawing punch with different surface characteristics, (a) Untreated tool (b) Air-blasted tool (c) Ion-irradiated tool

For the untreated tools, machined traces on the circumferential direction of punch surface are observed. While, for the air-blasted tools, the maximum height of the surface roughness is remarkably rough and the dispersion of the maximum height roughness is also large. In contrast, for the ion-irradiated tools, though machining marks can slightly be observed, peak of the asperities are removed by ion sputtering and smooth surface with no directional property can be obtained.

The tests were conducted under a nonlubricated condition. The drawing speed was 0.4mm/s and no blank holder force (BHF) was applied as mentioned. To evaluate the formability under each surface condition, punch force was measured with a micro-load cell. Additionally, to evaluate the drawn microcups, the cup surface roughness was measured by laser scanning confocal microscope (LEXT OLS-3000, Olympus Co.).

4.1.2. Punch load-stroke curves

In order to compare the punch force-stroke curve in each condition, the punch force is normalized. Normalized punch force, $\overline{P}$, during the deep drawing process can be described as

$$\overline{P} = P / (\pi \cdot d_p \cdot t_0 \cdot \tau_y) \qquad (1)$$

,where, d_P is punch diameter, t_0 is initial foil thickness, and τ_Y is shear yield stress of the blank material (Hu et al., 2007). The punch stroke is normalized with the drawing punch diameter, d_P. In this normalization, higher normalized force indicates the higher fraction of friction force to the whole forming force during the process. Fig.14 shows the comparative data of normalized punch load-stroke curves between the three tools of different surface asperities. The error bar of the curves indicates the standard deviation of the normalized punch force. Fig 15 summarizes the standard deviations of the maximum drawing and ironing force in each load-stroke curve.

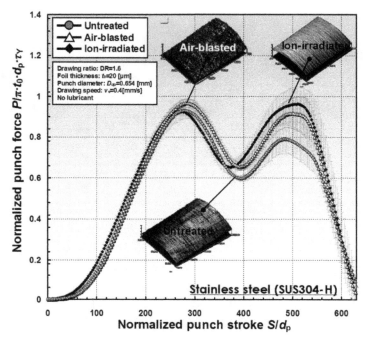

Figure 14. Comparison of normalized punch force-stroke curves between 3 tools with different surface properties

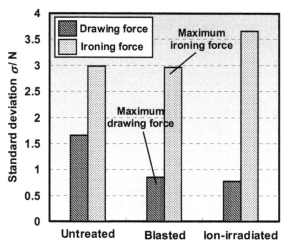

Figure 15. Comparison of standard deviation of maximum drawing and ironing force in each tool surface condition

The air-blasted tool shows the maximum drawing force, while the maximum ironing force is indicated by the ion irradiated tools. In addition, the both the lowest drawing and ironing force are shown by the untreated tool. In comparison of the dispersion of the force in each tool condition, the widest variation in maximum drawing force is observed for the untreated tool and the larger dispersion in maximum ironing force is indicated by the ion-irradiated tool. The wider variation of the ironing force data for every tool conditions seems to be attributed to the high contact pressure induced by the ironing process. Thus, the difference in the effect of tool surface properties on friction is markedly observed.

4.1.3. Surface quality of drawn cups

Fig.16 shows the surface images of the upper and the lower sides of the microcup wall of outer surface (Shimizu et al., 2009). The surface of the microcup has a large difference, as shown in the figure. In particular, for the microcup drawn by the ion-irradiated tool, scratches and chippings of the material caused by the adhesion are remarkable. In contrast, although the tool surface has the roughest surface for the air-blasted tool, the surface of the microcup is smoothened by sliding with the die, and better surface quality is observed. Furthermore, a roughened surface similar to an orange peel surface is observed at the lower part of the microcup drawn by the untreated tool. While, sliding traces with the die are observed under the conditions with the air-blasted and ion-irradiated tool.

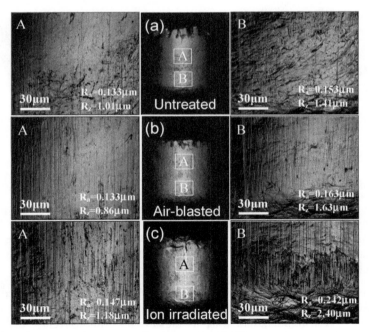

Figure 16. Surface images of microcup wall using 3 tools with different surface properties

The higher value and wider dispersion of the ironing force under the ion-irradiated tool appears to be due to the strong plowing of the wear particles. In fact, Yang et al. reported that WC particles would be exposed by the ion irradiation of WC-Co hard alloy, due to the difference in sputtering rate between the WC and Co (Yang et al., 2008). Since Co particles are removed from the surface, the WC particles lose the binding agents and are easily to drop off from the surface. Therefore the dropped WC particles seem to scratch and plough the surface of work material. Thus, remarkable difference in interfacial behaviour between the tool and material are experimentally demonstrated.

4.2. Effect of material surface properties

Secondly, the effect of material surface properties is studied. In order to compare the influence of material surface properties on micro-deep drawability, we focused on the aluminium foils, which have the two different surfaces in one foil sheet due to the manufacturing process of 2 layers rolling.

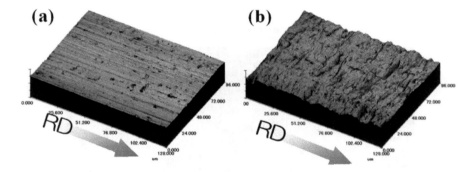

Figure 17. Surface images of 1N30 pure aluminum foil (a) bright surface, (b) mat surface

4.2.1. Materials and experimental conditions

As the aluminium foil, pure aluminium (JIS: 1N30) was used as test specimen. Two kinds of pure aluminium of spring-hard material (JIS:1N30-H) and annealed material (JIS: 1N30-O) were investigated. The nominal initial thickness of the both foils were 20μm. Table 3 shows the mechanical properties of the 1N30-H and the 1N30-O pure aluminium foil, which was provided by the supplier.

Fig.17 shows the surface images of the 1N30-H pure aluminium. The bright surface has smooth roughness of 1.02μmRz, while the mat surface has rough surface with 1.64μmRz. The contact between the bright surface and the die surface is defined as contact condition Br, and the condition for mat surface is defined as condition Mt.

Material	Young's modulus	Yield stress	Tensile strength	Elongation
1N30-H	70[GPa]	159[MPa]	177[MPa]	2.7[%]
1N30-O	70[GPa]	69[MPa]	88[MPa]	22.0[%]

Table 4. Mechanical properties of pure aluminium foils (JIS: 1N30, 20μm in thickness) used in the experiment.

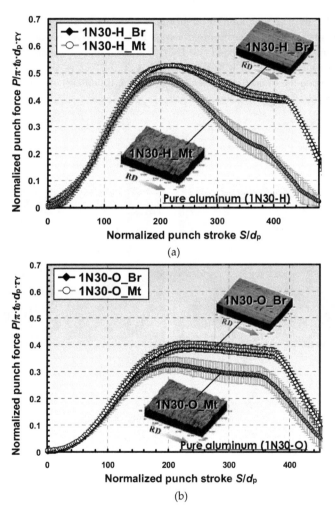

(a)

(b)

Figure 18. Punch force-stroke curves for comparing different material surface conditions (a) 1N30-H, (b) 1N30-O

The tests were carried out under the same conditions as previous section. Similarly, punch load during the process and surface quality of the cup after drawing were evaluated.

4.2.2. Punch load-stroke curves

Fig. 18 shows the normalized punch-stoke curves, which compares between different surface conditions of 2 kinds of pure aluminium foils, 1N30-H, and -O.

For both materials, the condition Br indicates the higher drawing and ironing force. Although the difference of maximum drawing force is almost no difference between condition Br and Mt for both 1N30-H and 1N30-O aluminium foil, maximum ironing force for the 1N30-H indicates larger difference than that of 1N30-O, as shown in Fig 18(a). In order to investigate the cause of these tendencies of each difference, the surface state of the drawn microcup is observed in following section.

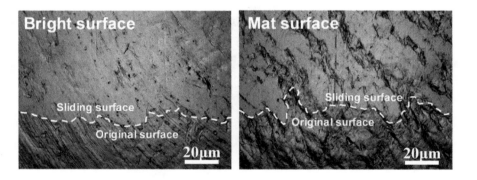

Figure 19. Surface images at bottom area of micro-drawn cup wall

4.2.3. Surface quality of drawn cups

Fig.19 compares the surface images of the bottom area of drawn cup wall surface under 2 contact conditions. As shown in the figure, the boundary of the sliding surface and the original surface with rolling traces can be clearly recognized. In comparison of the sliding surfaces, although almost of the whole area of the cup wall is smoothened with die surface for the bright surface condition, mat surface condition has the area, which does not contact with the die surface. This suggests that the real area of contact is less in condition Mt than that in condition Br during the ironing, and it would be responsible for the lower friction resistance during the forming. Therefore, it is clear that the initial material surface roughness is also responsible for the friction resistance during sheet metal forming.

4.3. FE analysis with surface roughness model

In order to discuss the obtained experimental results in detail, and to evaluate the experimental results more generally, FE analysis with proposed surface roughness model is carried out.

4.3.1. Simulation conditions

The simulation was carried out with the condition as mentioned in section 3.2. To study the effect of combination of surface geometry between the blank and the tools on formability, different combinations of surface geometries were analyzed. The combination conditions of the process are given in Table 5. To compare and quantify the effect of the surface topographical interaction between the tool and the material, the punch forces during the process were calculated.

		Blank surface geometry	
		$Rz=0.5\mu m$, P=5μm	$Rz=0.05\mu m$, P=5μm
Tool surface geometry	$Rz=0.5\mu m$ P=5μm	No.1 Blank Die	No.2 Blank Die
	$Rz=0.05\mu m$ P=5μm	No.3 Blank Die	No.4 Blank Die

Table 5. Combination conditions of surface geometry between blank and tools

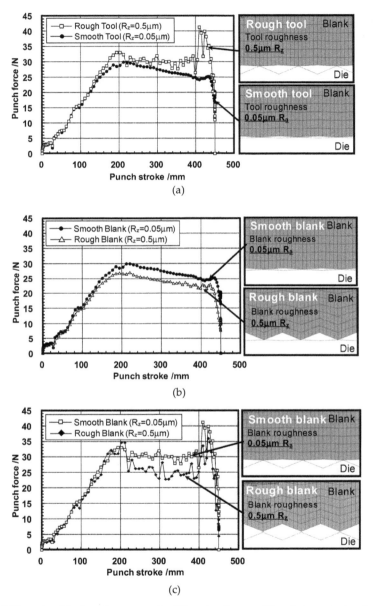

Figure 20. FEM results of load-stroke curve under different surface conditions between tool and material (a) Effect of tool surface amplitude (b) Effect of blank surface amplitude with smooth tool, (c) Effect of blank surface amplitude with rough tool

4.3.2. Simulation results

Fig.20 (a) shows the effect of roughness amplitude of a tool with same smooth material surface on formability (Condition No.2 and 4) (Shimizu et al., 2009). The curve obtained for the rough tool indicates the higher punch forces than those of the smooth tool. The difference in ironing force is particularly large. This tendency was also observed in the experimental results of the comparison between the untreated tool (smooth) and the air-blasted tool (rough), as shown in Fig. 14. From the deformation history in the analysis, it can be seen that, if the tool surface asperity are sharper than the blank surface asperity, the plastic deformation of the blank surface asperity would be easily occurred, due to the intensive surface pressure. In the actual contact behavior during the process, this phenomenon in FE analysis could be translated to the fracture of oxide film layer due to the high normal pressure. It would easily induce the adhesion or the plowing and increases the friction.

Fig. 20 (b) and (c) show the effect of blank roughness amplitude under the same tool surface condition (Shimizu et al., 2009). As for the condition with the smooth amplitude tool (Condition No.3 and 4) as shown in Fig. 20 (b), the punch force of the rough material is lower than that of the smooth material. The similar tendency in the experiment is already shown in Fig. 18(a) and (b). As mentioned from the observed surface images of micro-drawn cup in Fig.19, since the real area of contact under the smoother amplitude blank is much larger than that of rough amplitude blank, the friction force will increase and it results in the higher forming force.

While for the condition with a rough amplitude tool (Condition No.1 and 2) shown in Fig.20 (c), the maximum drawing and ironing force indicates almost the same value between the conditions. This shows the higher impact of the tool surface roughness than the material surface roughness. Since the harder tool surface asperities would plow the softer material surface, the material surface roughness seems to be less influence on the friction resistance under the contact with the rough tool surface.

Thus, local interfacial behavior in micro-deep drawing could be explained on the basis of classic theory of conventional tribology. However, a feature of the micro-scale region appears to be the higher sensitivity of the global forming behavior to the microscopic surface properties. Therefore, the proper surface design of tools and work materials in micro-scale becomes significantly more important than in conventional macro-scale.

5. Conclusion

From the perspective of the significant importance of tribological behaviour under dry friction in micro-sheet metal forming, this chapter has created an overview of the effect of surface topography of tools and workpieces on micro-sheet formability.

The advanced development of the micro-deep drawing experimental system has realized the investigation under the actual micro-scale range forming behaviour, such as the fabrication of tiny micro cups with minimum 0.5 mm in diameter. By providing the different microtools with different surface properties, the high sensitivity of the forming force to the different surface conditions of micro tools has been clearly recognized. Especially, the ironing force and its deviation have the large differences. The surface observation of the drawn cup demonstrates the importance of the tool surface roughness and the material compatibility between the tool and material, in view of the occurrence of adhesion and abrasive wear. In the additional investigation of the influence of material surface roughness, the higher friction force for the bright smooth surface is indicated. This appears to be attributed to the real area of contact during the process, which is strongly dependent on the initial surface geometry of the work material.

Furthermore, the FE model with surface asperities has been proposed and the availability and the validation of this model are demonstrated. The calculation results of this surface roughness model under the different combination conditions of surface asperities are well corresponded to the experimental results. The general tendency of the interaction of the surface asperities between the tools and the workpieces is explainable in terms of the classic theory of tribology.

Since the sensitivity to the surface properties for the global forming behaviour becomes higher with decreasing the scale dimensions, the proper design of the tool and the material are required more precisely in microforming. In other words, the microscopic geometry of the surface of tools and materials would become a significant parameter of controlling global deformation behaviour in microforming. The tribological optimisation by the surface texturing technology for micro-tooling and materials would be the future tasks.

Author details

Tetsuhide Shimizu, Ming Yang and Ken-ichi Manabe
Tokyo Metropolitan University, Japan

Acknowledgement

The authors gratefully acknowledge the support from JSPS (Japan Society for the Promotion of Science) under a JSPS Research Fellowships for Young Scientists. In the whole experimental work of micro-deep drawing, the authors also particularly indebted to Mr. Kuniyoshi Ito at Micro Fabrication Laboratory (formerly at Seki Corporation) for his valuable advice and support.

6. References

Aizawa, T., Itoh, K., Iwamura, E. (2010). Nano-laminated DLC Coating for Micro-Stamping, *Proceedings of the International Conference on Metal Forming, Steel Research International*, vol.81-9, (2010), pp.1169-1172, Toyohashi, Japan, September 19-22, 2010

Bhushan, B. (2002). Introduction to Tribology, Wiley, New York

Engel, U. (2006). Tribology in Microforming, *Wear*, Vol.260, (2006), pp. 265-273

Fuentes, G.G., Diaz de Cerio, M.J., Rodriguez, R., Avelar-Batista, J.C., Spain, E., Housden, J., Qin, Y. (2006). Investigation on the Sliding of Aluminium Thin Foils against PVD-coated Carbide Forming Tools during Micro-forming, *Journal of Materials Processing Technology*, vol.177, (2006), pp.644-648

Fujimoto, K., Yang, M., Hotta, M., Koyama, H., Nakano, S., Morikawa, K., Cairney, J.(2007). Fabrication of Dies in Micro-scale for Micro-sheet Metal Forming, *Journal of Materials Processing Technology*, vol.177, (2006), pp.639-643

Geiger, M.; Kleiner, M. ; Eckstein, R. ; Tiesler, N.& Engel, U. (2001). Microforming, *Annals of the CIRP - Manufacturing Technology*, Vol.50, Issue 2, (2001), pp. 445-462

Hanada, K., Zhang, L., Mayuzumi, M., Sano, T. (2003). Fabrication of Diamond Dies for Microforming, *Diamond and Related Materials*, vol.12, (2003), pp.757-761

Hu, Z. & Vollertsen, F. (2007); Tribological Size Effect in Sheet Metal Forming, *Proceedings of the International Conference on Tribology in Manufacturing Process, ICTMP2007*, pp.163-168, Yokohama, Japan, September 24-26, 2007

Krishnan, N., Cao, J., Dohda, K. (2007). Study of the Size Effect on Friction Conditions in Micro-Extrusion. Part 1. Micro-Extrusion Experiments and Analysis, *ASME Journal of Manufacturing Science and Engineering*, vol.129 No.4, (2007), pp.669–676

Manabe, K., Shimizu, T., Koyama, H., Yang, M., Ito, K. (2008). Validation of FE simulation based on surface roughness model in micro-deep drawing, *Journal of Materials Processing Technology*, vol. 204, (2008), pp.89-93

Mori, L., Krishnan, N., Cao, J., Espinosa, H. (2007). Study of the Size Effects and Friction Conditions in Micro-Extrusion. Part II. Size Effect in Dynamic Friction for Brass–Steel Pairs, *ASME Journal of Manufacturing Science and Engineering*, vol.129 No.4, (2007), pp.677–689

Putten, K.V., Franzke, M., Hirt, G. (2007). Size effect on friction and yielding in wire flat rolling, *Proceedings of the 2nd International Conference on New Forming Technology, ICNFT 2007*, pp. 583-592, Bremen, Germany, September 20-21, 2007

Shimizu, T., Murashige, Y., Ito, K., Manabe, K. (2009). Influence of surface topographical interaction between tool and material in micro-deep drawing, *Journal of Solid Mechanics and Materials Engineering*, Vol.3 No.2, (2009), pp.397-408

Vollertsen, F., Biermann, D., Hansen, H.N., Jawahir, I.S. & Kuzman, K. (2009). Size Effect in Manufacturing of Metallic Components, *Annals of the CIRP - Manufacturing Technology*, Vol.58, (2009), pp. 566-587

Vollertsen, F.& Hu, Z. (2006). Tribological Size Effects in Sheet Metal Forming Measured by a Strip Drawing Test, *Annals of the CIRP - Manufacturing Technology*, Vol.55, Issue 1, pp. 291-294

Vollertsen, F.& Hu, Z. (2008). Determination of Size-Dependent Friction Functions in Sheet Metal Forming with Respect to the Distribution of the Contact Pressure, *Production Engineering*, vol.2-4, (2008), pp.345–350

Yang, M. & Osako, A.(2008). Application of Ion Irradiation for Surface Finish of Micro-forming Die, *Journal of Materials Processing Technology*, vol. 201, (2008), pp.315-318

Design

Self-Consistent Homogenization Methods for Predicting Forming Limits of Sheet Metal

Javier W. Signorelli and María de los Angeles Bertinetti

Additional information is available at the end of the chapter

1. Introduction

Formability of sheet metals can be characterized by the forming-limit diagram (FLD). This concept has proved to be extremely useful for representing conditions leading to the onset of sheet necking (Hecker, 1975), and now is one of the best tools available to metallurgical engineers to assess a particular steel sheet's ability to be drawn or stretched. In a single diagram, the FLD represents all combinations of critical-limit surface strains corresponding to failure. Within the FLD, a line called forming-limit curve (FLC) separates the region of uniform sheet deformation from the region of slightly greater deformation, where the sheet will likely develop a local deformation instability or neck. Experimental measurement of the FLD is not an easy task, requiring a wide range of sample geometries and even more than one type of mechanical test. Also, many test factors measurably affect the limit-strain determination: friction conditions, small deviations in loading paths due bending effects, and strain-measurement procedures. Similarly, several physical factors related to material properties (e.g. plastic anisotropy, work hardening and strain-rate sensitivity) have an important influence in the development of localized necking or failure. Numerical simulation promotes a better understanding of deformation and failure in polycrystal sheet metal aggregates, by examining issues related to crystal anisotropy and stress / strain heterogeneity.

Considerable effort has recently been made to develop theoretical models for predicting the FLD behavior. Most of them are based either on a bifurcation analysis (Storen & Rice, 1975) or a model where the strain instability appears in the deformation process due to an imperfection already present in the material (Marciniak & Kuczynski, 1967). The latter, MK from now on, has probably been the most widely used of the two techniques. Within the MK framework, the influence of various constitutive features on FLDs has been explored using phenomenological plasticity models and crystal plasticity. In recent years, research has shown that the localization of plastic flow is influenced by deformation anisotropy (Asaro & Needleman, 1985; Tóth et al., 1996; Wu et al., 2004a; Lee & Wen, 2006). Thus, crystal-plasticity models should provide a framework for better understanding the relation between flow localization and material microstructure. Issues such as yield-surface shape – changes of sharpness – material anisotropy – crystal reorientation – are directly addressed within a polycrystalline model. It is widely recognized that the crystallographic texture strongly affects forming-limit diagrams and the macroscopic anisotropy of polycrystalline sheet metals. Numerous authors have adopted the MK model in conjunction with a crystal plasticity model to describe strain localization in rolled sheets (Kuroda & Tvergaard, 2000; Knockaert et al., 2002; Wu et al., 2004a; Inal et al., 2005; Yoshida et al., 2007; Neil & Agnew, 2009). Based on this strategy, the authors have examined how plastic anisotropy influences limit strains (Signorelli et al., 2009). For the FLD simulations, crystallographic effects were taken into account by combining the MK approach with a viscoplastic (VP) self-consistent (SC) and a Full-Constraint (FC) crystal-plasticity model, MK-VPSC and MK-FC respectively.

In this chapter we will analyze the influence that the numerous microstructural factors characterizing metals have on forming-limit strains. Moreover, we will focus on the consequences that selecting either a FC or SC type grain-interaction model has on numerical results. We will start, in the following section, with a brief description of the texture and anisotropy of cubic metals. The representation of crystallographic texture and the determination of the polycrystal texture are addressed. The material's plastic deformation as a result of crystallographic dislocation motion on the active slip systems is discussed at the end of the section. The single crystal properties and the way in which grains interact in a polycrystal are the subject of Section 3. An outline of the implementation of the VPSC formulation in conjunction with the well-known MK approach for modeling localized necking closes the section. A parametric analysis of the influence of the initial-imperfection intensity and orientation, strain-rate sensitivity and hardening on the limit strains is the content of Section 4. In Section 5 the MK-FC and MK-VPSC approaches will be examined in detail. FLDs will be predicted for different materials in order to clearly illustrate the differences between the FC or the VPSC homogenization schemes, particularly in biaxial stretching.

2. Texture and anisotropy of cubic metals

Plastic anisotropy of polycrystals arises from crystallographic texture. In a material with a plasticity-induced texture, anisotropy at the microscopic level is determined by the different ways in which the material is deformed. In metals, plastic deformation occurs by crystallographic slip, due to the movement of dislocations within the lattice. In general, slip takes place on the planes which possess the highest atomic density, *slip planes*, and in the most densely packed directions, *slip directions*. The slip plane is characterized by the unit vector **n**, which is normal to the plane, and the slip direction, represented by the unit vector **b** (Burger's vector). The combination of both vectors, which are perpendicular to each other, defines a *slip system*.

Since crystallographic slip is limited to certain planes and directions, the applied stress required to initiate plastic flow depends on the orientation of the stress relative to the crystallographic axis of the crystal. If the plane is either normal or parallel to the applied stress, the shear stress on the plane is zero and no plastic deformation is possible. Slip begins when the shear stress on a slip system reaches a critical value τ_c. This yield criterion is called *Schmid's Law*. In most crystals slip can occur either in the **b** or –**b** direction.

Figure 1 shows a slip system represented by the vectors **n** and **b**. Suppose that the crystal has a general state of stress σ_{ij} acting on it referenced to the coordinate system **S** (**S** is fixed to the sample). The shear stress σ'_{12} acting on the slip system can be obtained by transforming the stress tensor σ_{ij} from the **S** to the **S'** system (**S'** is fixed to the slip system). Using the typical equations for tensor transformation, the resolved shear stress acting on the slip system is:

$$\tau_r = \sigma'_{12} = b_i \; n_j \; \sigma_{ij} \tag{1}$$

If the crystal is loaded in tension along the X_3 axis, the shear stress acting on the slip plane is

$$\tau_r = \sigma \; cos \lambda \; cos \phi, \tag{2}$$

where λ is the angle between the slip direction and the tensile axis, and ϕ is the angle between the tensile axis and the normal to the slip plane.

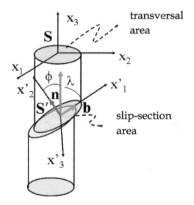

Figure 1. A schematic diagram of slip in the direction **b** occurring on a plane with the normal **n**.

In FCC materials, the crystallography of slip is simple, it takes place on the most densely packed planes {111} and in the most densely packed directions <110>. In BCC metals, the most common mode of deformation is {110}<111>, but these materials also slip on other planes: {112} and {123} with the same slip direction. Plastic deformation occurs by 12 crystallographic slip systems of the type {111}<110> for FCC metals and 48 slip systems of the type {110}<111>, {112}<111> and {123}<111> for the BCCs (see Table 1). A slip line is the result of a displacement of the material along a single lattice plane through a distance of about a thousand atomic diameters. The slip lines are visible traces of slip planes on the surface, and they can be observed when a metal with a polished surface is deformed plastically. As an example, in the optical micrograph shown in Figure 2, the slip bands appear as long steps on the surface. The terraced appearance is produced when the slip planes meet the crystal surface.

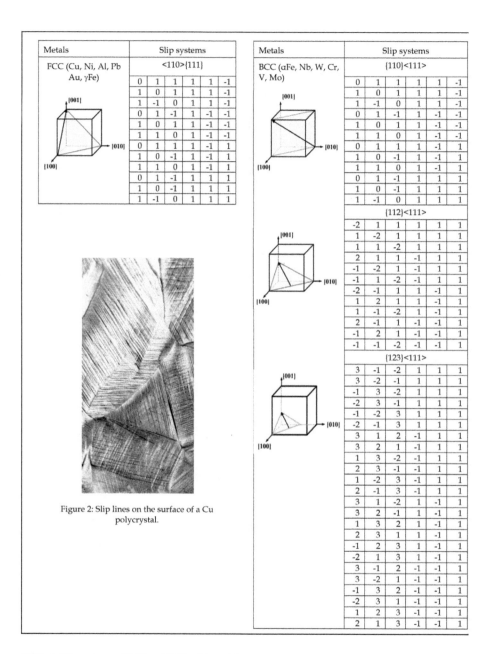

Metals	Slip systems					
FCC (Cu, Ni, Al, Pb Au, γFe)	<110>{111}					
	0	1	1	1	1	-1
	1	0	1	1	1	-1
	1	-1	0	1	1	-1
	0	1	-1	1	-1	-1
	1	0	1	1	-1	-1
	1	1	0	1	-1	-1
	0	1	1	1	-1	1
	1	0	-1	1	-1	1
	1	1	0	1	-1	1
	0	1	-1	1	1	1
	1	0	-1	1	1	1
	1	-1	0	1	1	1

Figure 2: Slip lines on the surface of a Cu polycrystal.

Metals	Slip systems					
BCC (αFe, Nb, W, Cr, V, Mo)	{110}<111>					
	0	1	1	1	1	-1
	1	0	1	1	1	-1
	1	-1	0	1	1	-1
	0	1	-1	1	-1	-1
	1	0	1	1	-1	-1
	1	1	0	1	-1	-1
	0	1	1	1	-1	1
	1	0	-1	1	-1	1
	1	1	0	1	-1	1
	0	1	-1	1	1	1
	1	0	-1	1	1	1
	1	-1	0	1	1	1
{112}<111>						
-2	1	1	1	1	1
1	-2	1	1	1	1
1	1	-2	1	1	1
2	1	1	-1	1	1
-1	-2	1	-1	1	1
-1	1	-2	-1	1	1
-2	-1	1	1	-1	1
1	2	1	1	-1	1
1	-1	-2	1	-1	1
2	-1	1	-1	-1	1
-1	2	1	-1	-1	1
-1	-1	-2	-1	-1	1
{123}<111>						
3	-1	-2	1	1	1
3	-2	-1	1	1	1
-1	3	-2	1	1	1
-2	3	-1	1	1	1
-1	-2	3	1	1	1
-2	-1	3	1	1	1
3	1	2	-1	1	1
3	2	1	-1	1	1
1	3	-2	-1	1	1
2	3	-1	-1	1	1
1	-2	3	-1	1	1
2	-1	3	-1	1	1
3	1	-2	1	-1	1
3	2	-1	1	-1	1
1	3	2	1	-1	1
2	3	1	1	-1	1
-1	2	3	1	-1	1
-2	1	3	1	-1	1
3	-1	2	-1	-1	1
3	-2	1	-1	-1	1
-1	3	2	-1	-1	1
-2	3	1	-1	-1	1
1	2	3	-1	-1	1
2	1	3	-1	-1	1

Table 1. Slip systems of FCC and BCC cubic metals.

2.1. Crystal orientation

A polycrystal is composed of crystals, each with a particular crystallographic orientation. Several parameters are involved in characterizing a polycrystal, such as the shape, size, crystallographic orientation and position of each grain inside the sample. The orientation of each crystal in the polycrystal can be defined by a rotation from the sample coordinate system to the crystal coordinate system. The sample coordinate system is referenced to the sample, and it can be chosen arbitrarily. For an example, the *Rolling Direction* (RD), the *Transverse Direction* (TD) and the *Normal Direction* (ND) are typically chosen as sample coordinate system for a rolled sheet. The orientation relation between a single crystal and the sample coordinate systems may be thought of as rotating one frame into the other. Euler angles are useful for describing one frame in term of the other, or vice versa. Several different notations have been used to define these angles, but that of Bunge is most common and will be used in this chapter (Bunge, 1982). These three angles represent three consecutive rotations that must be given to each grain to bring its crystallographic <100> axes into coincidence with the sample axes. This is equivalent to saying that any orientation can be obtained by conducting three elemental rotations (rotations around a single axis). Consequently, any rotation matrix can be decomposed into a product of three elemental rotation matrices. The matrix rotation (Eq. 3), written in terms of Euler angles $(\varphi_1,\phi,\varphi_2)$, is obtained by multiplication of the elementary matrices defining the three successive Euler rotations: i) a rotation about the Z-axis through the angle φ_1, ii) a rotation about the new X-axis through the angle ϕ and iii) a rotation about the last Z-axis through an angle φ_2. This gives the crystal coordinate system (see Figure 3).

$$\begin{pmatrix} \cos\varphi_1\cos\varphi_2 - \sin\varphi_1\sin\varphi_2\cos\phi & \sin\varphi_1\cos\varphi_2 + \cos\varphi_1\sin\varphi_2\cos\phi & \sin\varphi_2\sin\phi \\ -\cos\varphi_1\sin\varphi_2 - \sin\varphi_1\cos\varphi_2\cos\phi & -\sin\varphi_1\sin\varphi_2 + \cos\varphi_1\cos\varphi_2\cos\phi & \cos\varphi_2\sin\phi \\ \sin\varphi_1\sin\phi & -\cos\varphi_1\sin\phi & \cos\phi \end{pmatrix} \quad (3)$$

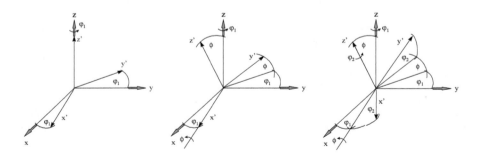

Figure 3. Definition and sequence of rotation through the different Euler angles.

2.2. Crystallographic texture

Texture refers to a non-uniform distribution of crystallographic orientations in a polycrystal. The textures of rolled or rolled and recrystallized sheets have been most widely investigated in metallurgy. Crystallographic orientations in rolled sheets are generally represented as being of the type {hkl}<uvw>, where {hkl} are the grain planes that lie parallel to the plane of the sheet. On the other hand, the <uvw> directions lie parallel to the rolling direction. Conventionally, the standard method of representing textures was by means of pole figures. However, while pole figures provide a useful description of texture, the information they contain is incomplete. A complete description can be obtained by the Orientation Distribution Function (ODF), which describes the orientation of all individual grains in the aggregate.

2.2.1. Pole figures

Texture measurements are used to determine the orientation distribution of crystalline grains in a polycrystalline sample. There are several experimental methods that can be used to measure texture. The most popular is X-ray diffraction. A pole figure – which is a projection that shows how the specified crystallographic directions of grains are distributed in the sample reference frame – results from an X-ray diffraction texture measurement. This representation must contain some reference directions that relate to the material itself. Generally these directions refer to the forming process.

The inverse pole figure is a particularly useful way to describe textures produced from deformation processes. In this case only a single axis needs to be specified. An inverse pole figure shows how the selected direction in the sample reference frame is distributed in the reference frame of the crystal. The frequency with which a particular crystallographic direction coincides with the sample axis is plotted in a single triangle of the stereographic projection.

2.2.2. ODF and Euler space

The ODF specifies the probability density for the occurrence of particular orientations in the Euler space. This space is defined by the three Euler angles, which are required to fully describe a single orientation. Mathematical methods have been developed that allow an ODF to be calculated from the numerical data obtained from several pole figures. The most widely adopted notations employed for these description were those proposed independently by Bunge and Roe. They used generalized spherical harmonic functions to represent crystallite distributions (A detailed description of the mathematics involved can be found in Bunge, 1982). ODF analysis was developed originally for materials with cubic crystallography and orthorhombic sample symmetry, i.e. for sheet products. In the Bunge notation, for cubic/orthotropic crystal/sample symmetry, a three dimensional orientation volume may be defined by using three orthogonal axes for φ_1, ϕ and φ_2, with each ranging from $0°$ to $90°$. The value of the orientation density at each point in Euler space is the strength or intensity of that orientation in multiples of random units.

Most of the texture data available in the literature and almost all of the ODF data refer to rolled materials. The information contained in a three-dimensional ODF can be expressed in terms of typical components and fibers for cubic symmetry materials. A fiber is a range of orientations limited to a single degree of freedom about a fixed axis, which appears as a line that may or may not lie entirely in one section of ODF. The ideal components and fibers are associated with more or less constant intensity for a group of orientations related to one another by rotations around a particular crystallographic direction.

During cold rolling of FCC metals, two crystallographic fibers arise: the α-fiber containing <110>//ND orientations and extending from *Goss* {110}<001> to *Brass* {110}<112>; and the β-fiber which starts at *Brass*, runs through *S* {123}<634>, and finally reaches *Copper* {112}<111>. The β-fiber contains the most stable components of the rolling texture (Humphreys & Hatherly, 2004). Considering recrystallized rather than rolled material, the typical texture components are *Cube* {001}<100> and *Goss*. Table 2 shows a schematic representation of the rolling texture characteristic of the {111} pole figure (left) and the main texture components for FCC (right). The nature of the FCC rolling texture is such that the data are best displayed in φ_2 sections, while the typical {100} and {111} pole figures best represent these orientations.

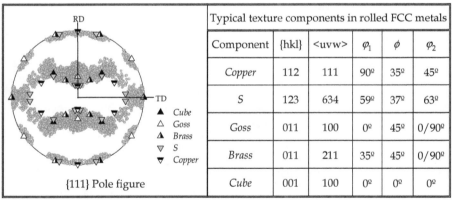

	Typical texture components in rolled FCC metals				
Component	{hkl}	<uvw>	φ_1	ϕ	φ_2
Copper	112	111	90º	35º	45º
S	123	634	59º	37º	63º
Goss	011	100	0º	45º	0/90º
Brass	011	211	35º	45º	0/90º
Cube	001	100	0º	0º	0º

{111} Pole figure

Table 2. FCC rolling components.

Cold rolling and recrystallization textures in BCC metals are commonly described in terms of five ideal orientations: {001}<110>, {112}<110>, {111}<110>, {111}<112> and {554}<225>. The positions of theses orientations in the {100} pole figure are shown at the left in Table 3. In general, BCC metals and alloys tend to form fiber textures. That is most orientations are assembled along two characteristic fibers that run through orientation space: the α-fiber and the γ-fiber. The RD or α-fiber runs from {001}<110> to {111}<110>, containing orientations with the <110> axis parallel to RD, and the γ-fiber runs from {111}<110> to {111}<112>, gathering orientations with a <111> axis parallel to ND. The two fibers intersect at the

{111}<110> component (Ray et al., 1994). The data are best displayed by sections at constant values of φ_1, but the most important texture features can all be found in the $\varphi_2 = 45^\circ$ section (right in Table 3). Table 3 also gives the Miller indices and Euler angles of the typical BCC texture components. The {100} and {110} pole figures best represent the ideal BCC material orientations.

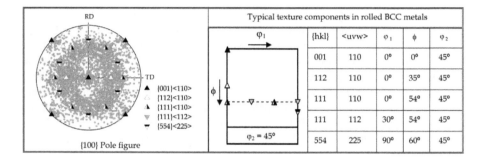

Typical texture components in rolled BCC metals				
{hkl}	<uvw>	φ_1	ϕ	φ_2
001	110	0°	0°	45°
112	110	0°	35°	45°
111	110	0°	54°	45°
111	112	30°	54°	45°
554	225	90°	60°	45°

Table 3. BCC rolling components.

3. Plasticity framework

We begin this section with the kinematic definitions of crystal-plasticity theory, citing the basic equations. The kinematic development of a single-crystal plasticity model has been well documented by several authors and is the subject of recent works (Kocks et al., 1998; Roters et al., 2010). Here we assume that, during plastic forming operations, it is possible to neglect the elastic contribution to deformation. Consequently, we will restrict ourselves to a rate-dependent plastic response at the single-crystal level.

3.1. Viscoplastic crystal plasticity

The velocity-gradient tensor, using a dot to indicate the time derivative, is given by

$$\mathbf{L} = \dot{\mathbf{F}}:\mathbf{F}^{-1} = \dot{\mathbf{R}}^*:\mathbf{R}^{*T} + \mathbf{R}^*:\mathbf{L}^p:\mathbf{R}^{*T}. \tag{4}$$

In this expression, $\mathbf{R}^*$ represents the crystallographic rotation, $\mathbf{F}$ corresponds to the effect of dislocation slip on the crystal deformation and $\mathbf{L}^p = \dot{\mathbf{F}}^p:\mathbf{F}^{p-1}$ is the plastic velocity gradient resulting from dislocation motion along specific planes and directions in the crystal (all potentially activated slip systems are labeled with the superscript s):

$$\mathbf{L}^p = \sum_s \left(\mathbf{n}^s \otimes \mathbf{b}^s \right) \dot{\gamma}^s, \tag{5}$$

here $\dot{\gamma}^s$ represents the dislocation slip rates, $\mathbf{n}^s$ and $\mathbf{b}^s$ are the normal to the system's or systems' glide plane and the Burgers' vector, respectively. They define the symmetric $\mathbf{m}^s$ and the screw-symmetric $\mathbf{q}^s$ parts of the Schmid orientation tensor:

$$\mathbf{m}^s = \frac{1}{2}\left(\mathbf{n}^s \otimes \mathbf{b}^s + \mathbf{n}^s \otimes \mathbf{b}^s \right), \tag{6}$$

$$\mathbf{q}^s = \frac{1}{2}\left(\mathbf{n}^s \otimes \mathbf{b}^s - \mathbf{n}^s \otimes \mathbf{b}^s \right). \tag{7}$$

The dislocation slip rates are derived using a viscoplastic exponential law (Hutchinson, 1976):

$$\dot{\gamma}^s = \dot{\gamma}_0 \left| \frac{\mathbf{m}^s : \mathbf{S}}{\tau_c^s} \right|^{1/m} sign(\mathbf{m}^s : \mathbf{S}). \tag{8}$$

where $\dot{\gamma}_0$ is the reference slip rate, τ_c^s is the critical resolved shear stress on the slip system labeled s, $\mathbf{S}$ is the deviatoric tensor stress and m is the strain-rate sensitivity exponent. The rate sensitivity m is typically quite small, a large value of $1/m$ tends to be almost a rate independent case, ~ 50. As $1/m \rightarrow \infty$, the plastic constitutive formulation becomes formally rate-independent.

The velocity gradient can be additively decomposed into symmetric and skew-symmetric parts

$$\mathbf{L} = \mathbf{D} + \mathbf{W}, \tag{9}$$

where $\mathbf{D}$ is the distortion rate tensor and $\mathbf{W}$ is the rotation rate tensor. They can be obtained by evaluating the symmetric and screw-symmetric parts of equation (4), respectively:

$$\mathbf{D} = \mathbf{R}^* : \mathbf{D}^\mathrm{p} : \mathbf{R}^{*\mathrm{T}}, \tag{10}$$

$$\mathbf{W} = \mathbf{\Omega} + \mathbf{R}^* : \mathbf{W}^\mathrm{p} : \mathbf{R}^{*\mathrm{T}}. \tag{11}$$

The rotation rate contains an extra contribution, the lattice spin tensor, defined as $\mathbf{\Omega} \equiv \dot{\mathbf{R}}^* : \mathbf{R}^{*\mathrm{T}}$. Rearranging Eq. (11) allows us to obtain the rate of change of the crystal orientation matrix:

$$\mathbf{\Omega} = \mathbf{W} - \mathbf{R}^* : \mathbf{W}^\mathrm{p} : \mathbf{R}^{*\mathrm{T}}, \tag{12}$$

which is used to determine the re-orientation of the crystal and consequently, to follow the texture evolution. The orientation change during plastic deformation can be described by a list of the Euler angle change rates, $\left(\dot{\varphi}_1, \dot{\phi}, \dot{\varphi}_2 \right)$, related to the lattice spin as follows:

$$\dot{\varphi}_1 = \Omega_{13} \frac{\sin \varphi_2}{\sin \phi} - \Omega_{23} \frac{\cos \varphi_2}{\sin \phi}$$

$$\dot{\phi} = -\Omega_{23} \cos \varphi_2 - \Omega_{13} \sin \varphi_2 \tag{13}$$

$$\dot{\varphi}_2 = \cos \phi \left(\Omega_{13} \frac{\cos \varphi_1}{\sin \phi} - \Omega_{23} \frac{\sin \varphi_1}{\sin \phi} \right) + \Omega_{21}$$

3.2. The 1-site VPSC-TGT formulation

For simulating the material response, a rate-dependent polycrystalline model is employed. In what follows, we present some features of the 1-site tangent VPSC-TGT formulation. For a more detailed description, the reader is referred to Lebensohn & Tomé (1993). This model is based on the viscoplastic behavior of a single crystal and uses a SC homogenization scheme for the transition to the polycrystal. Unlike the FC model, for which the local strain in each grain is considered to be equal to the macroscopic strain applied to the polycrystal, the SC formulation allows each grain to deform differently, according to its directional properties and the strength of the interaction between the grain and its surroundings. In this sense, each grain is in turn considered to be an ellipsoidal inclusion surrounded by a homogeneous effective medium, HEM, which has the average properties of the polycrystal. The interaction between the inclusion and the HEM is solved by means of the Eshelby formalism (Mura, 1987). The HEM properties are not known in advance; rather, they have to be calculated as the average of the individual grain behaviors, once a convergence is achieved. In what follows, we will only present the main equations of the VPSC model.

The deviatoric part of the viscoplastic constitutive behavior of the material at a local level is described by means of the non-linear rate-sensitivity equation:

$$\mathbf{D} = \dot{\gamma}_0 \sum_{s=1}^{\#sys} \mathbf{m}^s \frac{\mathbf{m}^s : \mathbf{S}}{\tau_c^s} \left| \frac{\mathbf{m}^s : \mathbf{S}}{\tau_c^s} \right|^{1/m - 1} = \mathbf{M} : \mathbf{S}, \tag{14}$$

where $\mathbf{M}$ is the visco-plastic grain compliance. The interaction equation, which relates the differences between the micro and the macro strain rates $(\mathbf{D}, \bar{\mathbf{D}})$ and deviatoric stresses $(\mathbf{S}, \bar{\mathbf{S}})$, can be written as follows:

$$\mathbf{D} - \bar{\mathbf{D}} = -\alpha \tilde{\mathbf{M}} : (\mathbf{S} - \bar{\mathbf{S}}) . \tag{15}$$

The interaction tensor $\tilde{\mathbf{M}}$, which is a function of the overall modulus and the shape and orientation of the ellipsoid that represents the embedded grain, is given by:

$$\tilde{\mathbf{M}} = \left(\mathbf{I} - \mathbf{S}^{esh} \right)^{-1} : \mathbf{S}^{esh} : \bar{\mathbf{M}} , \tag{16}$$

where $\mathbf{S}^{esh}$ is the Eshelby tensor; $\mathbf{I}$ is the 4th order identity tensor, and $\bar{\mathbf{M}}$ is the macroscopic visco-plastic compliance. The parameter α tunes the strength of the interaction tensor. In the present models, the standard TGT approach is used ($\alpha = 1$).

The macroscopic compliance can be adjusted iteratively using the following self-consistent equation:

$$\bar{M} = \langle M{:}B \rangle, \qquad B = \left(M + \tilde{M} \right)^{-1} : \left(\bar{M} + \tilde{M} \right), \tag{17}$$

where $\langle \; \rangle$ denotes a weighted average over all the grains in the polycrystal, and B is the accommodation tensor defined for each single crystal. The solution is reached using an iterative procedure that involves Eqs. (14), (15) and (17). It gives the stress in each crystal, the local compliance tensor and the corresponding polycrystal tensor, which is consistent with the impose boundary conditions.

3.3. Marciniak and Kuczynski technique

For simulating formability behavior, we implemented the VPSC formulation described above in conjunction with the well-known MK approach. As originally proposed, the analysis assumes the existence of a material imperfection such as a groove or a narrow band across the width of the sheet. In the approach's modified form, developed by Hutchinson & Neale (1978), an angle ψ_0 with respect to the principal axis determines the band's orientation (Fig. 4). Tensor components are taken with respect to the Cartesian X_i coordinate system, and quantities inside the band are denoted by the subscript b.

The thickness along the minimum section in the band is denoted as $h_b(t)$, with an initial value $h_b(0)$, while an imperfection factor f_0 is given by an initial thickness ratio inside and outside the band:

$$f_0 = \frac{h_b(0)}{h(0)}, \tag{18}$$

with $h(0)$ being the initial sheet thickness outside the groove.

Equilibrium and compatibility conditions must be fulfilled at the interface with the band. Following the formulation developed by Wu et al. (1997), the compatibility condition at the band interface is given in terms of the differences between the velocity gradients $\left(\bar{L}, \bar{L}^b \right)$ inside and outside the band respectively:

$$\bar{L}^b = \bar{L} + \dot{c} \otimes n. \tag{19}$$

Eq. (19) is decomposed into the symmetric, $\bar{D}$, and screw-symmetric, $\bar{W}$, parts:

$$\bar{D}^b = \bar{D} + \tfrac{1}{2}(\dot{c} \otimes n + n \otimes \dot{c}), \tag{20}$$

$$\bar{W}^b = \bar{W} + \tfrac{1}{2}(\dot{c} \otimes n - n \otimes \dot{c}) \tag{21}$$

Here, $\mathbf{n}$ is the unit normal to the band, and $\dot{\mathbf{c}}$ is a vector to be determined. The equilibrium conditions required at the band interface are given by

$$\mathbf{n}.\overline{\boldsymbol{\sigma}}^{b}\, h_{b} = \mathbf{n}.\overline{\boldsymbol{\sigma}}\, h, \tag{22}$$

where $\overline{\boldsymbol{\sigma}}$ denotes the Cauchy stress. Noting that δ_{ij} is the Kronecker symbol, the boundary condition $\overline{\sigma}_{33} = 0$ is applied as follows

$$\overline{\sigma}_{ij} = \overline{S}_{ij} - \overline{S}_{33}\, \delta_{ij} \qquad (i = 1,2,3). \tag{23}$$

The integration of the polycrystalline model inside and outside the band is performed in two steps. First, an increment of strain is applied to the material outside the band, $\overline{\mathbf{D}}\,\Delta t$, while the imposed strain path on the edges of the sheet is assumed to be

$$\rho = \frac{\overline{L}_{22}}{\overline{L}_{11}} = \frac{\overline{D}_{22}}{\overline{D}_{11}} = \text{const.} \tag{24}$$

It is assumed that $\overline{D}_{13} = \overline{D}_{23} = \overline{W}_{13} = \overline{W}_{23} = 0$ outside and inside the band. The instability appears in a narrow zone inclined at an angle ψ_0 with respect to the major strain axis. The equilibrium condition, Eq. (22), can be expressed in the set of axes referenced to the groove $\mathbf{n}, \mathbf{t}$ (see Fig. 4):

$$\begin{aligned} \overline{\sigma}_{nn}^{b}\, h_{b} &= \overline{\sigma}_{nn}\, h \\ \overline{\sigma}_{nt}^{b}\, h_{b} &= \overline{\sigma}_{nt}\, h. \end{aligned} \tag{25}$$

The compatibility condition requires equality of elongation in the direction $\mathbf{t}$,

$$\overline{D}_{tt}^{b} = \overline{D}_{tt}. \tag{26}$$

Because, we are considering thin sheets with the orthotropic symmetries in the plane of the sheet in this research, in-plane stretching results in a plane-stress state. As discussed by Kuroda & Tveergard (2000), when an orthotropic material is loaded along directions not aligned with the axes of orthotropy, it is necessary to compute the $\overline{L}_{12}$ component by imposing the requirement that $\overline{\sigma}_{12} = 0$. After solving each incremental step, the evolution of the groove orientation ψ is given by

$$\mathbf{n} = \frac{1}{\sqrt{t_1^2 + t_2^2}}\begin{pmatrix} -F_{11}\, t_1^0 - F_{12}\, t_2^0 \\ F_{21}\, t_1^0 + F_{22}\, t_2^0 \end{pmatrix}; \tag{27}$$

where $\mathbf{F}$ is the deformation gradient tensor.

The system of Eqs. (19) and (22, 23) can be solved to obtain $\dot{\mathbf{c}}$. This is done by substituting the macroscopic analogous Eq. (14) into the incremental form of Eq. (22) and using Eq. (20) to eliminate the strain increments in the band. At any increment of strain along the prescribed strain path, the non-linear system of two equations is solved (Signorelli et al., 2009). More recently, in Serenelli et al. (2011), Eqs. (25) and (26) were used after obtaining the state ($\bar{L}$, $\bar{\sigma}$) in the homogeneous zone, in order to solve the groove state avoiding the 2x2 set of non-linear equations mentioned above. In this case, the remaining unknowns $\bar{L}^b_{11}, \bar{L}^b_{12}, \bar{L}^b_{33}$ and $\bar{\sigma}^b_{22}, \bar{\sigma}^b_{13}, \bar{\sigma}^b_{23}$ are obtained by solving a mixed boundary-condition in the VPSC module, with the logic time benefits.

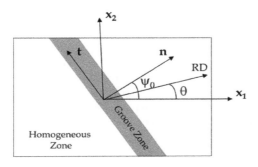

Figure 4. A thin sheet in the plane x_1-x_2 with an imperfection band.

To analyze the development of deformation localization during proportional straining, the calculations were performed over different strain paths. They were defined in terms of the strain-rate ratios $\rho = \bar{D}_{22} / \bar{D}_{11}$ over the range $-0.5 \le \rho \le 1$ (step = 0.1). The possible variations of the FLD for each ρ are obtained by performing calculations every 5 degrees of Ψ_0 to a maximum of 90 degrees. The failure strains ε^*_{11}, ε^*_{22} outside the band and the critical failure angle Ψ^* are obtained after minimizing the curve ε^*_{11} versus Ψ_0. In the present work, failure is assumed when $\left| \bar{D}^b_{33} \right| > 20 \left| \bar{D}_{33} \right|$.

4. Effects of the main model's parameters on the FLDs

In this section, we analyze the influences on the limit strains of the initial-imperfection intensity and orientation, the strain-rate sensitivity and the hardening to determine MK-VPSC performance. The sensitivity of the MK-VPSC model to the initial grain-shape and to texture and textural evolution is also addressed, by showing results obtained from sheets with rolling and random initial textures. In this section, the material inside and outside of the groove is taken to be a polycrystal described by 1000 equiaxed grains, except where noted otherwise. Each grain is assumed to be a single crystal with a FCC crystal structure. Plastic deformation occurs on 12 crystallographic slip systems of the type {111}<110>. We constructed the initial texture in both zones to be the same, and assumed a reference plastic shearing rate of $\dot{\gamma}_0$ = 0.001 s^{-1}. In order to account for the strain hardening between slip systems, we adopted isotropic hardening. In this case the evolution of the critical shear stresses is given by

$$\dot{\tau}_c = \sum_s h^s \left| \dot{\gamma}^s \right| \tag{28}$$

where h^s are the hardening moduli behaviors, which depend on Γ (accumulated sum of the single-slip contributions to γ^s). These moduli can be written using the initial hardening rate, h_0 , and the hardening exponent, n:

$$h^s = h_0 \left(\frac{h_0 \Gamma}{\tau_c^s n} + 1 \right)^{n-1} ; \qquad \Gamma = \sum_s \int_0^t \left| \dot{\gamma}^s \right| dt . \tag{29}$$

The strain-induced hardening law prescribed above is applied to all slip systems.

4.1. Initial imperfection

The MK approach predicts the FLD based on the growth of an initial imperfection. However, the strength of the imperfection cannot be directly measured by physical experiments. Zhou & Neale (1995) analytically predicted the effect of the initial imperfection parameter f_0 on the FLD and demonstrated, as expected, that the forming-limit strain decreases with increasing depth of the initial imperfection. Using the MK-VPSC approach, we too determined that the limit strains are greatly affected by the value of f_0. In addition, we found, as did Zhou & Neale, that the smaller the imperfection the larger the limit strain. The calculations plotted in Fig. 5 were performed using a random initial texture.

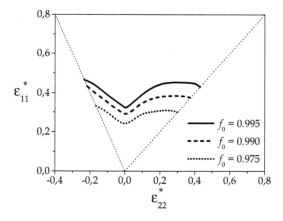

$(m = 0.01, n = 0.23, h_0 = 1410 \text{ MPa}, \tau_c^s = 47 \text{ MPa})$.

Figure 5. Influence of the initial imperfection f_0 on the FLD

All three curves show the minimum limit strain close to the in-plane plane-strain path. Although the limit strains are different for the different values of the initial imperfection, the profiles of the simulated cases are equivalent over the range $-0.5 \le \rho \le 0.3$. For $\rho > 0.3$, the profiles of the FLD curves remain insensitive to defect severity, except for paths approaching equal-biaxial deformation. In this case, the largest value of f_0 gives a more pronounced fall in the limit strain. Our results compare favorable to those of Wu et al. (1997), but these authors predict a more noticeable decrease of the limit-strains as the strain path approaches $\rho = 1.0$. It is possible that the FC homogenization scheme used by Wu et al., rather than the SC scheme of the present work, produced these differences. Also, we assumed a greater strain-rate sensitivity than did Wu and his co-authors. As we will see in the next section, larger rate sensitivities produce higher limit-strain profiles near $\rho = 1.0$.

4.2. Strain-rate sensitivity

The influence of the material's rate sensitivity m on the FLDs is addressed. Fig. 6 shows the calculated limit-strains assuming a random initial texture and an initial imperfection of $f_0 = 0.99$. Depending on the m value, not only are the limit strains different, but the FLD profiles vary as well. The limit-strains decrease with decreasing m, for m-values of 0.05, 0.02 and 0.01, while the FLD profiles in the negative minor-strain range ($\rho < 0$) continue to exhibit a nearly linear behavior. For biaxial paths ($\rho > 0$) near plane strain, the forming-limit strain increases rapidly for low values of m, while for high m, $m = 0.05$, the major limit-strain is nearly constant. As illustrated by the curves shown in Fig. 6, high values of m also displace the ending of the steep-slope profile towards an equi-biaxial path. Analyzing the results for $m = 0.01$ in the neighborhood of $\rho = 1.0$, we find decreasing limit-strains, indicating that lowering the strain-rate sensitivity produces a similar effect to that found using a Taylor model.

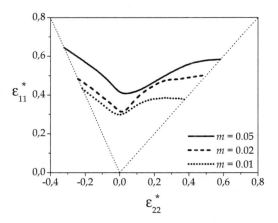

($f_0 = 0.99$, $n = 0.23$, $h_0 = 1410$ MPa, $\tau_c^s = 47$ MPa).

Figure 6. Influence of the rate sensitivity m on the FLD

The effects produced by the different strain-rate sensitivities are consistent with the relationship between the hardening behavior and m, described by Eqs. (28) and (29). The parameter m in the viscoplastic law controls the accumulated shear, which in turn drives the hardening. It is also known that as the m value increases textural sharpness decreases, though this behavior depends on the imposed strain path, too. The calculated average number of slip systems associated with the main strain paths and several additional material sensitivities are presented in Table 4. It is interesting to note that, as expected, the fewer the number of slip systems and the sharper the texture the lower the limit curve.

	$m = 0.05$	$m = 0.02$	$m = 0.01$
$\rho = -0.5$	4.5	2.8	2.3
$\rho = 0.0$	2.8	2.2	1.9
$\rho = 1.0$	3.3	3.1	2.8

Table 4. Calculated average number of active slip systems as a function of the deformation path and strain-rate sensitivity.

As opposed to the results for the limit-strain values, the critical angles at failure are almost insensitive to the strain-rate sensitivity. The predicted final angles rise to between $\Psi^* = 34°$-$37°$ for a uniaxial path and decrease to zero degrees for $\rho \geq 0$. If we restrict the model interaction to the FC hypothesis, as done by Wu et al. (1997), our calculated values equal theirs.

4.3. Hardening coefficients

Slip induced hardening is another important factor influencing the limit strains. From Eqs. (28) and (29) it is easy to see that the parameters h_0 and n govern the strain hardening. We investigated the effect produced by different values of the hardening coefficient n (0.16, 0.19 and 0.23) while fixing the other material properties. The calculated FLDs are shown in Fig. 7, where it is clear that the slip hardening coefficient n does not affect the shape of the forming limit curves. However, it can be seen that the largest value of n produces the highest limit strains. Also, no noticeable dependence with ρ is observed. Because we use isotropic hardening in these calculations, the n parameter only guides the stress level, producing these simple behaviors. This will not be the case when latent hardening and other kinematic effects are included in the calculations.

4.4. Grain shape

It is important for the reader to note that the VPSC model has the capability to account for the grain-shape and its evolution. The influence of different elongated grain-shapes on the limit strains is shown in Fig. 8 for a non-textured material. Generally, the calculated FLDs decrease when increasingly elongated grains are considered. In the negative minor-strain range, no appreciable changes are noted in the FLD's behavior for different aspect ratio grains. To the contrary, in the biaxial-stretching zone, differences are observed, especially when the aspect ratio is more pronounced. For initially equiaxed grains and grains elongated up to an aspect ratios of (3:1:1), the profiles of the simulated limit strains show no significant changes. However, for a (10:1:1) grain aspect ratio, we found a noticeable decrease of the limit-strains moving from plane-strain to equi-biaxial tension. This result can be correlated with the final textures predicted for each grain-shape morphology. The {111} pole figures calculated for the homogeneous deformation zone at the end of the equi-biaxial loading path are shown on the right in Fig. 8.

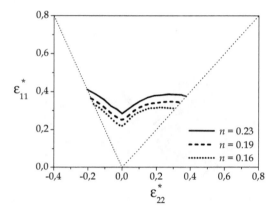

$(f_0 = 0.99, m = 0.01, h_0 = 1410 \text{ MPa}, \tau_c^s = 47 \text{ MPa}).$

Figure 7. Influence of the slip-induced hardening n on the FLD

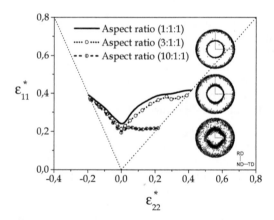

$(f_0 = 0.99, m = 0.02, n = 0.16, h_0 = 4000 \text{ MPa}, \tau_c^s = 22 \text{ MPa}).$

Figure 8. Influence of the initial grain shape on the limit-strains for a non-textured material

4.5. Effects of texture

A crystallographic texture develops during metal forming and it is a key component of the material's microstructure. It is generally accepted that this microstructural feature significantly affects forming-limit strains. To investigate these texture effects, we carried out simulations using two initial textures: one a random distribution of orientations (R) and the other a rolling texture (S). Two {111} pole figures illustrating these textures are shown in Fig. 9. To construct the S texture we first fixed the ideal component volume fractions, 10% {001}<100>, 15% {011}<100>, 30% {123}<634>, 10% {112}<111> and 35% {011}<211>, and then spread the distribution by assigning each grain a misorientation angle of $\theta < 15°$ with respect to the ideal component.

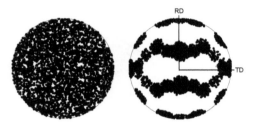

Figure 9. Grain orientation distributions represented by {111} pole figures: material R (left) and material S (right).

These different initial textures strongly affected the forming-limit curves as plotted in Fig. 10. In the negative minor-strain range ($\rho < 0$) of the FLD, the shapes are nearly straight lines with the maximum values at $\rho = -0.5$. The predictions, however, begin to diverge at $\rho = 0$, and the differences increase continuously, reaching a maximum for a biaxial deformation path. The S textured material develops a much stronger anisotropy than the R, likely producing the observed results. The S material's forming-limit curve slopes downwards from plane-strain to equi-biaxial tension, and over the whole range $\rho > 0$ the predicted forming limits for the R case are larger than those for the S. Fig. 10 includes a plot of the final textures of each sample at the end of the equi-biaxial loading path. Clearly, the R and S textures evolve to different states producing the strong effects observed in the FLD behavior.

Our calculations also show that the influence of crystallographic texture evolution is at least as important as effects of the initial grain distributions. Evolution effects have been previously discussed by several authors. Tóth et al. (1996) performed simulations with a rate independent Taylor model, showing that crystal rotations decrease the limit strains. Tang & Tai (2000), using the MK analysis together with continuum damage mechanics (CDM) and the Taylor model, found the same behavior for the limit strains. They claim that the development of texture causes deterioration of the material. On the other hand, Wu et al. (2004b) use a mesoscopic approach and a Taylor homogenization scheme to show that texture evolution increases the limit-strains in the biaxial zone. Finally, Inal et al. (2005) recently analyzed these two studies and their opposite conclusions, adding a study of how texture evolution in BCC materials affects the FLD. In the work of Inal et al., the simulations show that texture development does not have a significant influence on the FLD. Actually based on our simulations, we found it necessary to analyze the development of material anisotropy in a more complex way in order to determine the effect of texture updating on the limit strains. This in turn is captured more or less realistically by the different homogenization schemes.

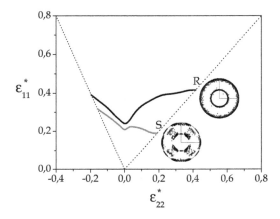

$(f_0 = 0.99, m = 0.02, n = 0.16, h_0 = 4000$ MPa, $\tau_c^s = 22$ MPa).

Figure 10. Influence of the initial texture on the FLD

Such an analysis can explain the opposite trends in limit strains reported by previous researchers, which cannot be understood based only on initial-material textures. In our opinion, the differences in the forming-limit strains are related to material anisotropy and its evolution along the deformation path. This produces an increase or decrease of the FLD profile. Many previous investigations have proven that the VPSC model gives a more realistic description of the anisotropic behavior of polycrystalline materials. We believe that results of the MK-VPSC strategy presented here are a better way to explain and justify the different effects of the material parameters.

To assess the influence of the texture evolution on the limit strains, we repeated the calculations shown in Fig. 10 but without texture evolution (Fig. 11). In negative strain space ($\rho < 0$) the FLDs have practically identical shapes, although the calculated values are slightly lower when the initial texture is not updated. However, in the biaxial zone, the tendency is quite different. The FLDs for both materials now approach each other, and a certain matching is observed. For the R texture, the limit-strain values reflect texture evolution. Texture and hence anisotropy evolution produces greater limit strains. In the case of the S material, when texture updating is off, the limit strains in the biaxial zone increase continuously, showing a different behavior than when the texture is updated. We attribute this behavior to the sharpness of the material yield locus and consequently, to the slip systems selected to accommodate the imposed deformation. The corresponding yield-loci after equi-biaxial stretching for both materials are displayed in Fig. 12 in $\sigma_{11} - \sigma_{22}$ space. Following Barlat's work (1989), the parameter p quantifies the effect of yield-surface shape on limit strains for the last four cases (Table 5). It should be noted that the trends of p and the predicted limit-strain values are consistent, as Hiwatashi et al. (1998) and Friedman & Pan (2000) have noted.

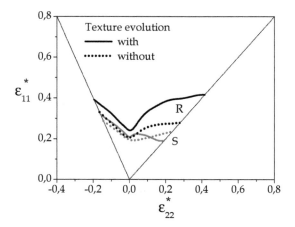

(f_0 = 0.99, m = 0.02, n = 0.16, h_0 = 4000 MPa, τ_c^s = 22 MPa).

Figure 11. Influence of texture evolution on the FLD

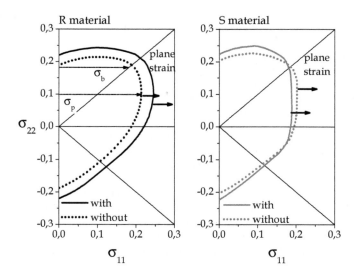

Figure 12. Calculated yield loci for materials R and S with (solid lines) and without (dotted lines) texture evolution. The equi work-rate surface is normalized to the work rate for uniaxial stretching, as calculated with the FC model.

Material	Texture evolution	p
R	Yes	1.159
S	No	1.072
R	Yes	1.131
S	No	1.087

Table 5. p-parameter for the R and S equi-biaxial stretching cases.

5. Discussions about MK–FC and MK–VPSC approaches

A comparison between the FC and the VPSC interaction models is the subject of this section. For this purpose, we calculate forming-limit strains using both homogenization schemes together with the MK approach. The numerical procedure, previously applied to a FCC structure, is extended to include the slip-system families of BCC polycrystals. The consequences of the FCC and BCC crystallographic-slip assumptions, coupled with the selection of either FC or SC type grain-interactions, are investigated in detail. Then, we focus on the effect of the cube texture on the forming-limit behavior, and seek to explain why a spread about cube exhibits unexpectedly high limit strains close to equi-biaxial stretching when the MK-FC is used. Finally, we explore the right-hand side of the FLD for a BCC material considering either 24 or 48 active slip systems for crystal-plasticity simulations. The advantages of using the VPSC material model in the MK approach are discussed at the end of this section. For all calculations in this section, and as was pointed out in Section 4, the strain hardening between slip systems is taken into account by adopting isotropic hardening.

In what follows, we apply the MK-VPSC and MK-FC approaches for predicting FLDs to both FCC and BCC materials. We assume that plastic deformation occurs by 12 crystallographic slip systems of the type {111} <110> for the FCC material and 48 slip systems of the type {110} <111>, {112} <111> and {123} <111> for the BCC. The crystal level properties listed in Table 6 are determined by imposing the same uniaxial behaviors for all simulations. An initially random texture, described by 1000 equiaxed grains is assumed. The slip resistances τ_c^s of all slip systems are taken equal, the rate sensitivity is $m = 0.02$, and a reference slip rate of $\dot{\gamma}_0 = 0.001$ s^{-1} is assumed.

Material	FCC-FC	FCC-SC	BCC-FC	BCC-SC
h_0 (MPa)	1950	2720	1850	3100
n	0.250	0.224	0.250	0.265
τ_c^s (MPa)	31.5	47.0	37.0	45.0

Table 6. Material parameters used in the simulations.

To analyze the development of deformation localization during proportional straining, the calculations are performed assuming an initial imperfection of $f_0 = 0.99$ over the different strain paths. The predicted limit strains are presented in Fig. 13. For each homogenization method, both materials have about the same profile from uniaxial tension ($\rho = -0.5$) to in-plane plane-strain tension ($\rho = 0$). Over this entire range, the major limit strains decrease with increasing ρ. It can be seen that, for the BCC-SC material, the largest value of the parameter n results in the highest limit values. Also, it is interesting to note that MK-FC simulations using the same n value predict similar limit strains for the in-plane plane strain condition.

Predictions begin to diverge in the biaxial-stretching range. Here, results clearly illustrate large differences between the homogenization schemes and between materials. These differences reach a maximum for the equi-biaxial deformation path. The MK-FC framework predicts both the highest and lowest limit strains, for the BCC and FCC materials respectively. The MK-FC FCC material calculation leads to a remarkably low limit curve. Completely the opposite behavior is observed within the MK-VPSC scheme. In this case, the FCC material shows better formability than the BCC for $\rho \geq 0.3$, and for both materials, the calculated limit-strain curves remain between those calculated with the MK-FC scheme near the equi-biaxial zone. In the case of the FCC material, the MK-VPSC approach predicts a noticeable increase of the limit strains over the whole right side of the diagram, while the BCC material only shows that behavior in the region $0 \leq \rho \leq 0.6$. For $\rho \geq 0.6$, the MK-VPSC limit-strain values are nearly constant. Fig. 13 also includes the {100} stereographic pole figures of each material at the end of the equi-biaxial stretching path. As can be seen, FCC and BCC material textures evolve differently depending on the model assumption. For a FCC material, the FC model develops a weaker texture than that produced by the VPSC calculation. For the BCC case, the final textures are qualitatively similar but quantitatively different in their degree of intensity.

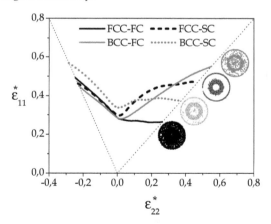

Figure 13. Influence of the slip microstructure and interaction model on the FLD.

According to Lian et al. (1989), the yield-surface shape has a tremendous effect on the FLD, and Neale & Chater (1980) demonstrated that a decrease in the sharpness of the stress potentials in equi-biaxial stretching promotes larger limit strains. A sharp curvature allows the material to quickly select a deformation path approaching plane strain, and this results in the prediction of a relatively low limit strain. The yield potentials of the materials were calculated by imposing different plastic strain-rate tensors under a state of plane stress in $\sigma_{11} - \sigma_{22}$ space. With the simulation, we deformed the material in equi-biaxial stretching up to a given plastic strain and then performed the yield-locus calculations. For the purpose of comparing polycrystal yield surfaces, all work-rates were normalized to that of FCC-FC uniaxial stretching. We calculated the yield loci corresponding to each of the necking limit strains, in order to highlight the link between the yield-surface shape and the forming-limit behavior. Shapes and curvatures predicted by the FC and VPSC models are shown in Fig. 14.

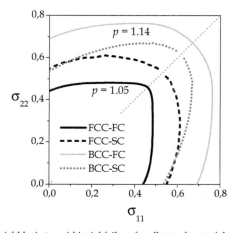

Figure 14. Calculated yield loci at equi-biaxial failure for all tested materials.

As expected, the curvature of the yield locus of the FCC-FC material in the equi-biaxial stretching zone is much sharper than those of the other materials, which is consistent with its limit-strain value, lowest among the cases considered, shown in Fig. 13. Similarly, the curvature of the BCC-FC yield locus is rounder than the others, which is again in agreement with the predicted limit strains. We also calculated the parameter p for our materials finding that the values of p for the FCC-FC and BCC-FC materials are the lowest and the highest, respectively.

The FLD results depend on the homogenization scheme, and the differences are explained in terms of the sharpness of the yield-loci and texture development. The MK-FC framework predicts both the highest and lowest limit strains, for the BCC and FCC materials respectively.

5.1. Influence of cube texture on sheet-metal formability

As mentioned in section 2.2, in rolled FCC sheets, crystallographic textures are frequently classified in terms of the ideal rolling and recrystallization components. Such classifications are well suited for theoretical modeling where mathematical descriptions of particular components can be input into simulations. In particular, we focus on how the strength of the cube texture affects localized necking. To investigate this effect, we modeled variations of the cube texture. The variations were constructed with different spreads of grain orientations around the ideal cube component. The procedure for modeling textures is the same as that used in Signorelli & Bertinetti (2009). As an example, the cube-15° texture is one whose grains have a misorientation with respect to the ideal cube orientation {100}<001> of less than 15°, uniformly distributed over that area. Fig. 15 shows the {111} stereographic pole figures for cube-3°, cube-7°, cube-11° and cube-15° distributions. For the cube set of textures, the number of individual orientations was set in order to obtain an adequate representation of a uniform distribution.

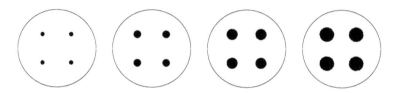

Figure 15. {111} pole figures used in the simulations for cube-3°, cube-7°, cube-11° and cube-15° distributions.

In the following FLD simulations, in both the homogeneous and MK band zones, standard FCC {111}<110> crystal slip is used, and the initial textures are assumed to be the same. Fig. 16 shows the predicted limit strains for the cube set of texture distributions using the MK-FC and MK-VPSC approaches. The simulations clearly show that there are differences between the VPSC and FC homogenization schemes, although the shapes and levels of the predicted FLDs are similar in the uniaxial range. No significant differences were found between the ideal cube and the cube-3° textures, since for this case the HEM nearly corresponds that of a single crystal. In these cases, both models closely predict shape and tendency. In the negative minor-strain range ($\rho < 0$) of the FLD, the shapes are nearly straight lines with the maximum values at $\rho = -0.5$. For both textures, the FLD curves slope downwards from plane-strain tension to equi-biaxial stretching, with the minimum limit-strain values at $\rho = 1$. These values are far below that of the random texture.

We would expect that a spread around the ideal orientation would give greater formability and higher limit strains. The FLD calculated from the random texture should be above all others, with the FLDs of particular spreads lying between those of the ideal cube and the random cases. Wu et al.'s (2004a) results do not show this expected behavior; their calculated forming-limit curves for cube-11° and cube-15° are significantly higher than that of the random texture near equi-biaxial stretching. These results were confirmed by Yoshida et al. (2007) using the same modeling hypothesis. Our simulations are similar to those reported by Wu et al. when the MK-FC approach is used. To the contrary, FLDs calculated with the MK-VPSC approach behave as expected. For $0 < \rho \leq 1$, the limit strains move upwards with an increase in the dispersion cut-off angle, and the strain-limit values never reach those of the random case.

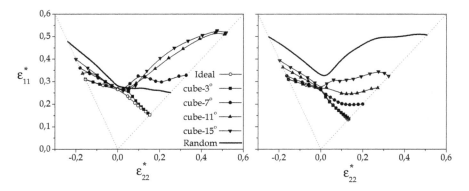

$(f_0 = 0.99,\ m = 0.02,\ n = 0.24,\ h_0 = 1218\ \text{MPa},\ \tau_c^s = 42\ \text{MPa})$.

Figure 16. Calculated FLDs for MK-FC (left) and MK-VPSC (right) models

In order to assess the effect of the yield-surface shape on the forming-limit behavior close to the balanced-biaxial stretching zone, $\rho = 1.0$, we prestrained the texture sets along the equi-biaxial path. The amounts of equi-biaxial strain corresponded to the necking-limit strains. Then, we calculate the yield-loci for cube-11°, cube-15° and random cases, using FC and VPSC models. The corresponding $\sigma_{11} - \sigma_{22}$ projections are shown in Fig. 17. The equi work-rate surfaces are normalized to the work rate for uniaxial stretching as calculated with the FC model. As expected, the yield loci are quite different. The curvatures of the VPSC yield loci are blunter than those of FC model, particularly for the random texture. This explains the higher limit-strains predicted by the MK-VPSC model as shown in Fig. 16. For the cube-11° and cube-15° initial textures, the FC yield loci are sharper and larger. As other researchers concluded and our simulations confirm, regions of reduced yield-locus curvature correspond to lower FLD values.

Fig. 18 shows the initial and final (at failure) inverse pole figures of the cube-15º for both constitutive-model approaches at $\rho = 1.0$. We found that the behavior of certain crystallographic orientations depends on the interaction model used. Particularly, near the <100> orientation, results of the models diverge. Using the VPSC approach, no grains remain close to <100> ($\Theta < 5º$), but for the FC simulations this is not the case, and the grains rotate in widely different directions. In both cases, one can trace an imaginary line that delineates a zone containing a high density of orientations and one vacant of orientations. The grain orientations tend to rotate and accumulate in the region approximately defined by <115> - <114> and <104> - <102> for FC and by <116> - <115> and <104> - <305> for VPSC, respectively. In addition, we found that the FC final orientations are distributed rather uniformly in the inhabited region. Interestingly, for the VPSC calculations, there is a preference to rotate half-way up the <104> - <305> segment line. Similar behavior can be found for the cube-11º distribution.

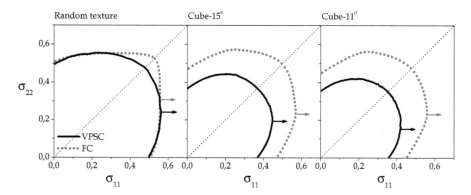

Figure 17. Calculated yield loci for VPSC and FC models. The equi work-rate surfaces are normalized to the work rate for uniaxial stretching, as calculated with the FC model.

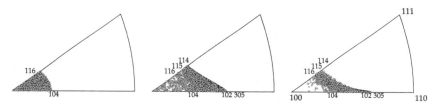

Figure 18. Stereographic triangles showing the initial cube-15° texture (left) and predicted final orientations after equi-biaxial stretching to failure by FC (center) and VPSC (right) models.

Simulations of FLDs show that the MK-FC strategy leads to unrealistic results, since an increasing spread about the cube texture produces unexpectedly high limit strains. However, results with the MK-VPSC approach successfully predict a smooth transition in the limit strains from the ideal-cube texture, through dispersions around the cube texture with increasing cut-off angles, ending with a random texture.

5.2. Influence of the dislocation slip assumption on the formability of BCC sheet metals

The identification of the active slip systems is a widely discussed issue in the plastic deformation of BCC crystals. The most common deformation mode is {110}<111>, but BCC materials also slip on other planes, {112} and {123}, with the same slip direction. In the literature, it is common for two sets of possible slip systems describing BCC plastic behavior to be considered: {110}<111>, {112}<111> (BCC24); or {110}<111>, {112}<111>, {123}<111> (BCC48). In what follows, we explore the right-hand side of the FLD for a BCC material using the proposed MK-FC and MK-VPSC approaches, and test the crystallographic slip assumption.

The crystal level properties are determined, by imposing same uniaxial behaviors for all the cases: BCC48-FC, BCC24-FC, BCC48-SC and BCC24-SC. Accordingly, the hardening parameters are chosen to give an identical uniaxial-stress response. They are listed in Table 7. The initial texture, the reference plastic shearing rate and shear strain-rate sensitivity are the same as used in the previous section.

Material	BCC48-FC	BCC48-SC	BCC24-FC	BCC24-SC
h_0 (MPa)	808	808	795	980
n	0.23	0.26	0.23	0.26
τ_c^s (MPa)	30.5	40.0	30.0	37.0

Table 7. Material parameters used in the simulations.

The predicted limit strains are shown in Fig. 19. Large differences are between the MK-FC and MK-VPSC results, particularly near equi-biaxial stretching, regardless of the material. For each homogenization method, both materials have about the same forming limit in plane strain. At $\rho = 0$, we found no difference in the predicted limit-strains values given by either the BCC24 or BCC48 approach, but a discrepancy appears between the FC and SC polycrystal models. Within the MK-VPSC framework, the profiles of the BCC48 and BCC24 simulations are very close for the strain ratios $\rho \leq 0.6$, and differences can be seen near equi-biaxial stretching. The MK-FC limit strains are similar as ρ increases up to 0.8, but for $\rho > 0.8$, the critical values calculated with the BCC24 deformation model increase to an unrealistic high value at $\rho = 1$. Our simulations clearly show that in equi-biaxial stretching the exclusion of {123}<111> crystallographic slip as a potential active deformation mode promotes higher limit strains for both models. For MK-VPSC this gap is only an increase in the limit strain from 0.37 to 0.39; whereas for MK-FC the value changes from 0.53 to 0.97.

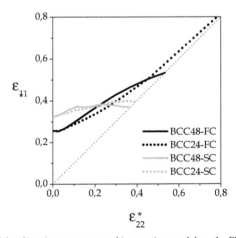

Figure 19. Influence of the slip microstructure and interaction model on the FLD.

The yield potentials of the materials were calculated by imposing different plastic strain-rate tensors under a state of plane stress in the $\sigma_{11} - \sigma_{22}$ section. With the simulation, we deformed the material in equi-biaxial stretching up to a given plastic strain and then performed the yield-locus calculations. For the purpose of comparing polycrystal yield surfaces, all work-rates were normalized to the case of BCC48-FC uniaxial stretching. We calculated the yield loci corresponding to each of the necking limit strains, in order to highlight the effect of the yield-surface shape on the forming-limit behavior. Particular attention must be paid to the surface's curvature near the balanced-biaxial stretching zone.

The shapes and curvatures predicted by the FC and VPSC models are quite different (Fig. 20 left). Within the VPSC framework, the yield loci are sharper, and only small differences can be found between the BCC48 and BCC24 based simulations. This is consistent with the similar limit-strain values predicted by the MK-VPSC model, as shown in Fig. 19. The differences are more pronounced for the MK-FC calculations. The curvature of the BCC24 yield locus is more gradual than that of the BCC48 material, again in agreement with the predicted limit strains. Fig. 20b presents the stress potential in a different but qualitatively similar way, based on the direction of plastic-strain rate and loading direction. In the Cauchy stress reference frame, the directions at different points along the predicted yield surfaces are characterized by two angles, θ and φ, as shown in Fig. 20 (right). These angles are taken to be zero along the horizontal axis and assumed positive in a counterclockwise sense. Differences in the sharpness of the stress potentials in equi-biaxial stretching, reflected in the slope of the plots, are clearly illustrated. For the BCC24-FC case, the direction of the plastic-strain rate $\bar{\mathbf{D}}$, or θ, seems nearly invariant in the vicinity of $\varphi=45^\circ$ (equi-biaxial strain-rate states), whereas for BCC48-FC the values of θ increase steadily with φ over this range. The curves calculated with the FC theory, in accordance with the observed yield loci, are not steep and clearly different for the two materials. For the BCC48-SC and BCC24-SC materials, the slight changes observed in the predicted critical-strain values correlate to almost identical profiles in Fig. 20b. In the vicinity of $\varphi=45^\circ$, the sharpness of both yield loci are characterized by an abrupt change of θ. Over the range $42^\circ \le \varphi \le 47^\circ$, θ varies linearly from 23° to 63°. This allows the material to quickly approach a plane-strain state with minor variations of stress state. We verified that this behavior can be mainly attributed to the ability of the VPSC model to distribute the imposed deformation according to the relative hardness of the grains. In addition, we note that a large majority of the grains, approximately 80 percent, shows a similar local-states solution (i.e. stress / strain-rate states, plastic dissipation and accumulated shear), but built from different sets of active slip systems when the simulations were carried out with either BCC48 or BCC24 slip assumptions. Consequently, MK-VPSC predicts very similar limit strains for both microstructural slip assumptions, though this result is specifically dependent on the initial material texture.

In summary, although it is normally accepted that a BCC material can be represented using 24 or 48 slip systems, we found that the MK-FC FLD calculations are sensitive to the material plasticity assumption in the vicinity of equi-biaxial stretching. However, this is not the case when we follow the MK-VPSC procedure, at least for a nontextured material. For these calculations, we found that either BCC24 or BCC48 materials give similar FLD curves, over the whole range of deformations.

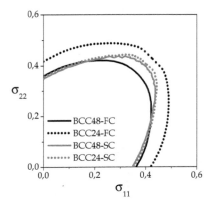
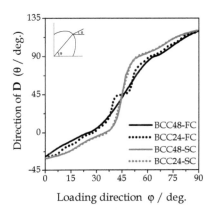

Figure 20. Yield loci at equi-biaxial failure for all tested materials (left); directions of the plastic strain-rate vectors (right).

Finally, we evaluated the MK-VPSC capability comparing the predictions of our model with recently published experimental and numerical results. Data and predictions for two low-carbon steel, LCS, sheets are analyzed. The limit-strain calculations performed with MK-FC and MK-VPSC are analyzed and discussed in terms of the crystallographic- slip assumption, and compared with the measured data. All experimental tests were conducted at room temperature.

For a first verification, experimental data are taken from Serenelli et al. (2010). The initial texture of the steel sheet was measured by using a Phillips X'Pert X-ray diffractometer. Incomplete pole figures for the {110}, {200} and {112} diffraction peaks were obtained. From these data, the ODF was determined, and the completed {110} and {100} pole figures were calculated (Fig. 21). The texture was discretized into 1000 orientations of equal volume fraction. Due to the annealing process, the texture contains a slight α-fiber ({001}<110> 1.2%, {112}<110> 9.6%), a more intense γ-fiber ({111}<110> 18.7%, {111}<112> 9.0%), and {554}<225> 11.7%. Optical micrographs of the LCS annealed microstructure show oblate spherical grains, with an approximate aspect ratio of 1.0:1.0:0.4.

Equi-biaxial bulge tests with different elliptical die rings were conducted in order to obtain three biaxial paths in strain space. The die ring masks had a major diameter of 125 mm and aspect ratios of 0.5, 0.7 and 1.0. The die rings and a corresponding tested circle-gridded specimens are shown in Fig. 22. A grid pattern of 2.5 mm diameter circles was electro-chemically etched on the surface of specimens. Details of the experimental procedure can be found in Serenelli et al. (2010).

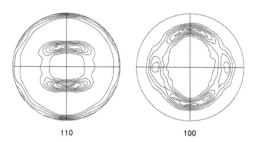

110   100

Figure 21. Experimental {110} and {100} stereographic pole figures (lines are multiples of a half-random distribution). Reference frame: X_1 top (RD), X_2 right (TD), X_3 center (ND).

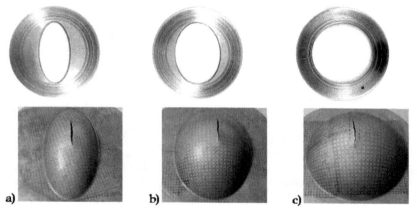

a)   b)   c)

Figure 22. Die-ring masks (top) and photograph of the gridded specimens after the tests (bottom) with ratios of 0.5 (a), 0.7 (b) and 1.0 (c) between the major and minor diameters. The marked points where the grid was measured can be observed in the photographs.

The alloy's hardening parameters were estimated in order to fit tensile test data. The coefficients for the BCC48-FC, BCC24-FC, BCC48-SC and BCC24-SC materials are listed in Table 8.

Material	BCC48-FC	BCC48-SC	BCC24-FC	BCC24-SC
h_0 (MPa)	1680	2900	1770	2900
n	0.203	0.212	0.201	0.212
τ_c^s (MPa)	49.0	60.0	47.5	59.0

Table 8. Material parameters used in the simulations for the LCS sheet.

The initial value of the imperfection factor, f_0, was taken to be 0.996. The FLD predictions are shown in Fig. 23 together with the bulge-test data. The BCC48-SC predicted limit-strains agree well with the measured points. The plane strain behavior is similar to that predicted for an initial non-textured material, and no influence from the crystallographic slip assumption is found. However, the differences between the interaction models remain. In these calculations, the MK-VPSC results showed a sensibility to the addition of {123}<111> slip as a potentially active deformation mode; only BCC48-SC predicts a realistic strain-limit profile.

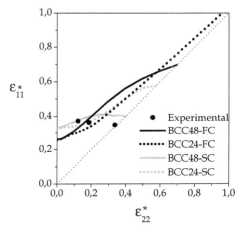

Figure 23. Influence of the slip microstructure and interaction model on limit strains for the LCS rolled sheet.

For a second verification, we consider data – experimental FLD and material's properties – from Signorelli et al. (2012) for an electro-galvanized DQ-type steel sheet 0.67 mm thick. Texture measurements were conducted using X-Ray diffraction in a Phillips X'Pert Pro-MPD system equipped with a texture goniometer, CuK alpha radiation and an X-ray lens. The initial pole figures obtained for the {110}, {112} and {100} diffraction peaks are shown in Fig. 24 (left). From these data the ODF was calculated. The measured texture represented by the $\varphi_2 = 45°$ section is also presented in Fig. 24 (right). It shows a high concentration of orientations with {111} planes lying parallel to the sample (sheet) surface together with the {554}<225> orientations. This is typical of a cold-rolled and annealed steel.

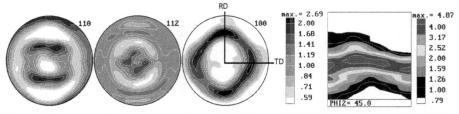

Figure 24. Experimental equal-area pole figures {110}, {112} and {100} (left); $\varphi_2 = 45°$ section of the ODF (right).

The forming-limit diagrams were determined by following an experimental procedure involving three stages: applying a circle grid to the samples, punch stretching to maximum load, and measuring strains. As we are not focused on the experimental methods and techniques, we will not present experimental details here. Readers are referred to Signorelli et al. (2012) for a completed description of the specific techniques for measuring the FLDs.

Simulations were performed following the methodology described in previous sections. The measured initial texture was discretized into 1000 orientations of equal volume fraction. In this case, we assumed that plastic deformation occurred by slip on the {110}, {112} and {123} planes with a <111> slip direction for each case (BCC48). The hardening parameters were established by numerically fitting the uniaxial tensile data taken parallel to the rolling direction with the following results: τ_c^s = 62 MPa, h_0 = 2275 MPa and n = 0.222 for VPSC simulations; and τ_c^s = 55 MPa, h_0 = 1100 MPa and n = 0.209 for the FC calculations. For the calculations, the initial slip resistances, τ_c^s, of all slip systems are assumed equal, the strain-rate sensitivity and the reference slip rate at the crystal level were taken to be m = 0.02 and $\dot{\gamma}_0^s$ = 1 s^{-1}, respectively. The simulated and the experimental curves are shown in Fig. 25.

Before performing simulations, we adjusted f_0 such that the predicted limit strains matched the experimental results in plane strain. For the MK-FC and MK-VPSC simulations these values of f_0 were 0.999 and 0.996 respectively. Together with the experimental data, the simulated FLDs for both the MK-FC and MK-VPSC schemes are shown in Fig. 26. The open symbols define a safe zone of uniform deformation for metal forming. The solid symbols correspond to measured circles that experienced local necking or fracture, specifying an insecure zone for metal forming.

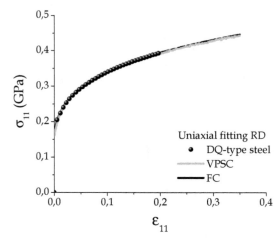

Figure 25. Experimental and simulated uniaxial tests parallel to the rolling direction.

The shapes and levels of the predicted FLDs for both models are similar in the tension-compression range, and the trends between measured and simulated limit strains are close, except near uniaxial tension. In this region, simulations show that there are differences between the MK-VPSC and MK-FC schemes. MK-FC predictions are more conservative and this curve lies below the region of localized flow. Examining the calculated FLDs in the biaxial quadrant of strain space, we found that the limit values predicted by MK-VPSC model accurately separate the regions of safe (uniform) and insecure (localized) deformation. On the other hand, the critical values calculated with the MK-FC approach, are only accurate for strain-path values to 0.3. These differences reach a maximum for the equi-biaxial deformation path. Our simulations clearly show that limit values calculated with the MK-FC approach increase unrealistically as ρ increases. It is clear that the VPSC scheme together with the MK approach provides accurate predictions of the DQ-type steel behavior over the entire biaxial range.

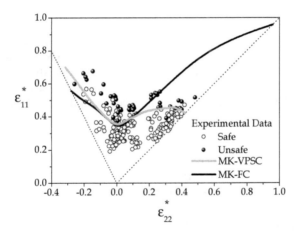

Figure 26. Experimental data and simulated FLDs for both MK-FC and MK-VPSC schemes for the DQ-type steel.

5.3. Summary

From the results presented in this chapter it can generally be concluded that the calculation of the FLD is strongly influenced by the selected constitutive description. In the present work we highlight the important role that the assumed homogenization scheme plays, which cannot be omitted in the discussion of the simulation's results. The predicative capability of a particular crystal-plasticity model is then assessed by comparing its predictions with those experimental data not used for the fitting. In our case, the discussion is framed in terms of the predicted FLD, texture evolution and polycrystal yield surface.

The emphasis in this chapter has been on cubic metals, and all calculations were carried out using either Full-Contraint or Self-Consistent models. All simulations clearly show that there are differences between the MK-VPSC and MK-FC assumptions, although the shapes and levels of the predicted FLDs are similar in the tension-compression range. Some examples were analyzed in order to highlight these differences:

- Non-textured FCC and BCC materials were investigated, imposing the same uniaxial behaviors for all simulations. The FLDs clearly depended on the homogenization scheme, and those differences were interpreted in terms of the sharpness of the yield-loci and texture development.
- For FCC texture materials, the MK-VPSC approach successfully predicts a smooth transition in the limit strains from the ideal cube texture, through dispersions around the cube texture with increasing cut-off angles, ending with a random texture. Important differences that increase with an increasing spread in texture distribution, particularly near equibiaxial strain-path, were found between the MK-FC and the MK-VPSC models.
- In order to verify the capability of the MK-VPSC model for predicting actual experimental limit strains for BCC textured materials, we carried out simulations for two different LCS steel sheets. The simulations gave good predictions of the steel's behavior over the complete biaxial range. To the contrary, the MK-FC model predicts extremely high limit strains as pure biaxial tension is approached, though both MK-FC and MK-VPSC predictions appear to be accurate on the tensile side of the plane strain.

Author details

Javier W. Signorelli and María de los Angeles Bertinetti
Instituto de Física Rosario (IFIR), CONICET–UNR, Rosario, Argentina

Acknowledgement

The authors thank M. G. Stout for fruitful comments about the present manuscript.

6. References

Asaro, R. & Needleman A. (1985). Texture development and strain hardening in rate dependent polycrystals. *Acta Metallurgica* 33, 923–53.

Barlat, F. (1989). Forming limit diagrams-predictions based on some microstructural aspects of materials. In: *Forming Limit Diagrams: Concepts, Methods and Applications*, Wagoner, R.H., Chan, K.S., Keeler, S.P. (Eds.), 275–302, The Minerals, Metals and Materials Society, Warrendale.

Bunge, H.J. (1982). *Texture Analysis in Materials Science—Mathematical Methods*, (2nd. Ed.), Butterworth, ISBN 0408106425, 9780408106429, London.

Friedman, P.A. & Pan, J. (2000). Effects of plastic anisotropy and yield criteria on prediction of forming limit curves. *International Journal of Mechanical Sciences* 42, 29-48.

Hecker, S.S. (1975). Simple technique for determining FLC. *Sheet Metal Industries* 52, 671-675.

Hiwatashi, S., Van Bael, A., Van Houtte, P. & Teodosiu, C. (1998). Predictions of forming limit strains under strain path changes: applications of an anisotropic model based on texture and dislocation structure. *International Journal of Plasticity* 14, 647-669.

Humphreys, F.J. & Hatherly, M. (2004). *Recrystallization and Related Annealing Phenomena*, (2nd. Ed.), Elsevier Science Ltd., Oxford OX5 IGB, ISBN-10: 008042, UK.

Hutchinson, J.W. (1976). Bound and self-consistent estimated for creep of polycrystalline materials. *Proceedings of the Royal Society of London* A 348, 101-127.

Hutchinson, J.W. & Neale, K.W. (1978). Sheet necking II, time-independent behavior. In: *Mechanics of Sheet Metal Forming*, Koistinen, D.P., Wang, N.M. (Eds.), 127-153, Plenum Press, New York, London,.

Inal, K., Neale, K. & Aboutajeddine, A. (2005). Forming limit comparison for FCC and BCC sheets. *International Journal of Plasticity* 21, 1255-66.

Knockaert, R., Chastel, Y. & Massoni, E. (2002). Forming limits prediction using rate-independent polycrystalline plasticity. *International Journal of Plasticity* 18, 231-47.

Kocks, U.F., Tomé, C.N. & Wenk, H.-R. (1988). *Texture and Anisotropy: Preferred Orientations in Polycrystals and Their Effect on Materials Properties*, Cambridge University Press, ISBN 052179420X, 9780521794206,UK.

Kuroda, M. & Tvergaard,V. (2000). Forming limit diagrams for anisotropic metal sheets with different yield criteria. *International Journal of Solids and Structures* 37, 5037-59.

Lebensohn, R.A. & Tomé C.N. (1993). A self-consistent approach for the simulation of plastic deformation and texture development of polycrystals: application to Zr alloys. *Acta Metallurgica et Materialia* 41, 2611-2624.

Lee, W.B. & Wen, X.Y. (2006). A dislocation-model of forming limit prediction in the biaxial stretching of sheet metals. *International Journal of the Mechanical Sciences* 48, 134-44.

Lian, J., Barlat, F. & Baudelet, B. (1989). Plastic behavior and stretchability of sheet metals. II. Effect of yield surface shape on sheet forming limit. *International Journal of Plasticity* 5, 131-147.

Marciniak, Z. & Kuczynski, K. (1967). Limit strains in the process of stretch-forming sheet metal. *International Journal of the Mechanical Sciences* 9, 609–20.

Mura, T. (1987). *Micromechanics of Defects in Solids*. Martinus Nijhoff Publishers, Dordrecht, ISBN 9024732565, 9789024732562, The Netherlands.

Neale, K.W. & Chater, E. (1980). Limit strain predictions for strain-rate sensitive anisotropic sheets, *International Journal of the Mechanical Sciences* 22, 563-574.

Neil, J.C. & Agnew, S.R. (2009). Crystal plasticity-based forming limit prediction for non-cubic metals: Application to Mg alloy AZ31B. *International Journal of Plasticity* 25 (3), 379-98.

Ray, R.K., Jonas, J.J. & Hook, R. E. (1994). Cold rolling and annealing textures in low carbon and extra low carbon steels, *International Materials Reviews* 39 (4), 129-172.

Roters, F., Eisenlohr, P., Bieler, T. & Raabe, D. (2010). Introduction to Crystalline Anisotropy and the Crystal Plasticity Finite Element Method, In: *Materials Science and Engineering*, Wiley-Vch Verlag GmbH & Co. KGaA, ISBN: 978-3-527-32447-7, Weinheim.

Serenelli, M.J., Bertinetti, M.A. & Signorelli, J.W. (2010). Investigation of the Dislocation Slip Assumption on Formability of BCC Sheet Metals. *International Journal of the Mechanical Sciences* 52, 1723-1734.

Serenelli, M.J., Bertinetti, M.A. & Signorelli, J.W. (2011). Study of limit strains for FCC and BCC sheet metal using polycrystal plasticity. *International Journal of Solids and Structures* 48, 1109-1119.

Signorelli, J.W., Bertinetti, M.A. & Turner, P.A. (2009). Predictions of forming limit diagrams using a rate-dependent polycrystal self-consistent plasticity model. *International Journal of Plasticity* 25 (1), 1-25.

Signorelli, J.W. & Bertinetti, M.A. (2009). On the Role of Constitutive Model in the Forming Limit of FCC Sheet Metal with Cube Orientations. *International Journal of the Mechanical Sciences* 51 (6), 473-480.

Signorelli, J.W., Bertinetti, M.A. & Serenelli, M.J. (2012). Experimental and numerical study of the role of crystallographic texture on the formability of an electro-galvanized steel sheet. *Journal of Materials Processing Technology* 212, 1367-1376.

Storen, S. & Rice, J.R. (1975). Localized necking in thin sheet. *Journal of the Mechanics and Physics of Solids* 23, 421–41.

Tang, C.Y. & Tai, W.H. (2000). Material damage and forming limits of textured sheet metals. *Journal of Materials Processing Technology* 99, 135-140.

Tóth, L., Dudzinski, D. & Molinari, A. (1996). Forming limit predictions with the perturbation method using stress potential functions of polycrystal visco- plasticity. *International Journal of the Mechanical Sciences* 38, 805-24.

Wu P.D., Neale K.W. & Giessen V.D. (1997). On crystal plasticity FLD analysis. *Proceedings of the Royal Society of London* A 453, 1831-48.

Wu, P.D., MacEwen, S.R., Lloyd, D.J. & Neale, K.W. (2004a). Effect of cube texture on sheet metal formability. *Materials Science and Engineering* A 364, 182-7.

Wu, P.D., MacEwen, S.R., Lloyd, D.J. & Neale, K.W. (2004b). A mesoscopic approach for predicting sheet metal formability. *Modelling and Simulation in Materials Science and Engineering* 12, 511-527.

Wu, P.D., Graf, A., Mac Ewen, S.R., Lloyd, D.J., Jain, M. & Neale, K.W., 2005. On forming limit stress diagram analysis. *International Journal of Solids and Structures* 42, 2225-2241.

Yoshida, K., Ishizaka, T., Kuroda, M. & Ikawa, S. (2007). The effects of texture on formability of aluminum alloys sheets. *Acta Materialia* 55, 4499-506.

Zhou, Y. & Neale, K.W. (1995). Predictions of forming limit diagrams using a rate-sensitive crystal plasticity model. *International Journal of the Mechanical Sciences* 37, 1-20.

Chapter 8

The Design of a Programmable Metal Forming Press and Its Ram Motion

Weizhong Guo and Feng Gao

Additional information is available at the end of the chapter

1. Introduction

Metal forming is one of the oldest production processes that can be traced back to the beginning of the Industry Revolution. Today, metal forming press is still one of the most commonly used manufacturing machineries. Every day, millions of parts are produced by metal forming ranging from battery caps to automotive body panels. Therefore, even a small improvement may add to significant corporative gain.

In the early days, metal forming presses used a simple slide crank mechanism that converted the rotating motion and the energy from the unprogrammable motor to the linear motion of the ram to form the workpiece. Throughout the years, the study of the metal forming reveals the desirable press performance: smooth pressing to avoid large transient force and vibration, long dwelling time to ensure uniform metal deformation, and slow releasing to minimize workpiece spring-back.

To achieve the desirable performance, many different types of metal forming presses have been developed. In general, they can be divided into two types: mechanical presses and hydraulic presses. The former is fast (high speed presses may reach up to several thousand shots per minute) and energy efficient (the large flywheel eases the impulsive force), but lacks flexibility. The latter is flexible (their motions, including the travel and the velocity, can be programmed), but is expensive to build and to operate. This is because the hydraulic system is expensive and must be constantly pressurized during the operation with or without the metal forming work. Recently, there are programmable mechanical presses driven by servomotors. They are a newest version of the mechanical presses that is programmable and of better perfromance, e.g. low noises, excellent efficiencies and high precision for metal forming operations such as drawing, stamping, blanking, etc. Compared with traditional mechanical presses, programmable mechanical presses provide less stamping tonnage due to the capability limit of servomotors.

The programmable mechanical presses are also evolved. Since early in the nineteen nineties, several press manufacturers have been developing a couple of servo mechanical presses in series. Komatsu [1] developed servo presses whose punching forces are from 35 tons up to 200 tons. Amada [2] produced SDE series so-called digital AC servo press with a capacity of 45 tons up to 300 tons. Amino [3] developed servo press with a punching force up to 2500 tons. Chinfong [4] produces servo presses with punching forces between 80 tons and 260 tons. Heavy-duty servo mechanical presses require heavy-duty servomotors. The cost of a heavy-duty servomotor increases at exponential growth with nominal power and torque, and the capacity of the servomotor available has to be constrainted by the development of the heavy-duty servomotor technology. Currently, it is reported that the servomotors have their capacities up to 200 KW that only few manufaturers can make. Using two 200 KW servomotors, Amino developed one largest servo mechanical press with a punching force of 2500 tons in the world [3]. With the servo mechanical press growing in tonnage, it is more difficult to find such a heavy-duty servo motor at a relatively low cost, even no such heavy-duty servomotors are available. Besides the researches sponsored by the machine tool industry, some theoretical and experimental researches were performed in the universities around the world. In [5], a servomotor-driven multi-action press was designed that connected the ram to a servomotor using a bolt-screw mechanism. The desired ram motion could be generated by programming the motion of the servomotor. A double-knucle servo mechanical press with 30ton punching force was developed in [6]. In [7] a slider-crank type servo punching machine was built.

To achieve a larger tonnage, a new solution was proposed in [8-10] based on hybrid machine concept [11, 12] (see Figure 1) and developed a prototype with 25 ton (see Figure 2). This hybrid press is actuated by one common motor and one servomotor through a 2-Degree-Of-Freedom (2-DOF) linkage that the servomotor was expected to contribute only a fraction of the total energy. Currently, the hybrid press is still under development and a couple of problems need to be solved before pratical application.

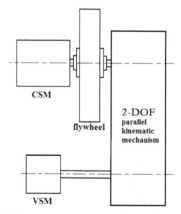

Figure 1. The conception of a hybrid press

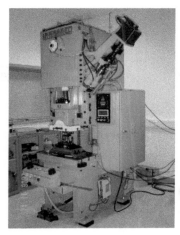

Figure 2. The prototype of a hybrid press

Generally, the development of a servo mechanical press mainly involves programmable actuator and control technology, punching linkage design, and ram motion design technology. For current designs, servomotors are normally used as programmable actuators. To develop heavy-duty programmable mechanical presses at relatively low cost, a new design is proposed and being developed in our laboratory to achieve higher tonnage by using more than one servomotor to drive the 1-DOF punching linkage [13]. This new scheme is called redundant actuation, or composite actuator. It provides an optional solution to heavy-duty actuation at a relatively low cost or at a higher power than single servomotor available. To combine two or more programmable actuators into a composite actuator, a 2-DOF or multi-DOF parallel mechanism is a feasible solution from a mechanical point of view (see Figure 3), similar to a yoke used between a pair of or more oxen to allow them to pull a load. In [14], several schemes of redundant actuation were proposed based on parallel mechanisms (see Figure 4). Apparently, a composite actuator should be able to act as one servomotor of large capability in terms of nominal torque, nominal power, nominal speed, and efficiency. To function as one servomotor, the mechanism to combine the programmable actuators should be of high efficiency, of high mechanical advantage and of high structural stiffness. It is obvious that a composite actuator can accommadate a slim un-synchronization of individual servomotors and even fault-tolerant of individual servomotors' disorder. Therefore, a slight difference between the motions of the two servomotors is permitted by the mechanical structure of the servo press to let the ram move freely.

In this chapter, the evolution history of the metal forming presses is outlined first and then redundant actuation scheme is discussed. The optimized design of the punching mechanism is detailed. Also, the ram motion is optimized using pseudo-NURBS expression for the stamping operation. Finally, case study is given, and a prototype with 200 t punching force is fabricated in the laboratory. The presented work provides an optional solution to the development of heavy-duty servo mechanical presses.

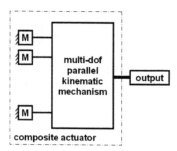

Figure 3. The composite actuator scheme

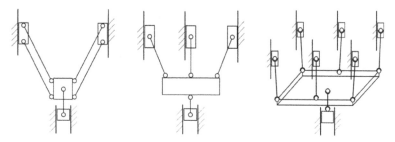

(a) 2-DOF mechanism   (b) 3-DOF mechanism   (c) 6-DOF mechanism

Figure 4. Some configurations for composite actuator design

2. The design of a new programmable mechanical press

In our laboratory, a servo press prototype with redundant actuation is fabricated with nominal punching force of 200 t. The design will be detailed in following sections.

2.1. The redundant actuation scheme

As mentioned above, two or more servomotors can be combined into a composite actuator through a 2-DOF or multi-DOF parallel mechanism. As shown in Figure 5, the presented servo mechanical press with redundant actuation uses two servomotors denoted as M to drive one input shaft which then drives a punching mechanism. These two servomotors are combined through a 2-DOF screw mechanism. To produce a high accuracy of the stamping operation, a host/slave control mode is adopted in the controller design of the servo press that the motion of one servomotor acting as host is followed by the motion of the other servomotor serving as slave. Hence, the two DOFs in the redundant actuation module avoid the bias of the ram and interference of the two servomotors while the host/slave control between the two dofs ensures high accuracy of the stamping operation. Therefore, this new scheme can provide a higher capacity with two servomotors or have a lower cost than current servo press.

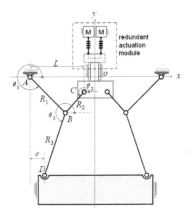

Figure 5. The schematics of the servo press with redundant actuation of 200 tonnage

2.2. The mechanism design in the redundant actuation

As discussed above, the redundant actuation scheme uses a 2-DOF screw mechanism to combine two servomotors as input of the punching mechanism. Herein, the mechanism design is one of the key problems for successful development. To design the 2-DOF mechanism, reasonable performance indices and the dimensional design should be addressed.

2.2.1. The performance indexing

Generally, a machine design relates to a couple of performance indices. As to the composite actuator for the servo press, the mechanism should have a high mechanical advantage to lower the torque required from servomotors for the stamping operation. Therefore, the mechanical advantage is an important performance index for the composite actuator design. For the two servomotors, it is straightforward to use mechanical advantage reciprocals of the mechanism as performance indices derived according to theoretical mechanics as follows:

$$\eta_{ma1} = \frac{M_{in1}}{F_{out}} = \frac{L_1}{2\pi} \cdot \frac{\cos\theta_1 \sin(\phi_2 + \theta_2 - \delta)}{\cos\delta \sin(\phi_2 + \theta_2 - \phi_1 - \theta_1)} \tag{1}$$

$$\eta_{ma2} = \frac{M_{in2}}{F_{out}} = \frac{L_2}{2\pi} \cdot \frac{\cos\theta_2 \sin(\phi_1 + \theta_1 - \delta)}{\cos\delta \sin(\phi_2 + \theta_2 - \phi_1 - \theta_1)} \tag{2}$$

$$\eta_{ma} = \frac{M_{in1} + M_{in2}}{F_{out}} = \frac{L_1 \cos\theta_1 \sin(\phi_2 + \theta_2 - \delta) + L_2 \cos\theta_2 \sin(\phi_1 + \theta_1 - \delta)}{2\pi \cos\delta \sin(\phi_2 + \theta_2 - \phi_1 - \theta_1)} \tag{3}$$

where, M_{in1}, M_{in2} and F_{out} are two input torques and output force of the mechanism respectively. As shown in Figure 6, L_1 and L_2 are leads of the two screw rods between the servomotors and the mechanism respectively, ϕ_1 and ϕ_2 are angles measured from vertical lines to two screw/slide axes respectively, δ is the angle measured from vertical line to link l_4, and θ_1 and θ_2 are angles measured from screw/slide axes to links l_1 and l_2 respectively. Obviously, θ_1 and θ_2 are functions of motion inputs of the servomotors.

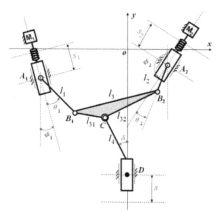

Figure 6. The general case for the composite actuator design

From Eq.(1), it is easy to know that the driving torques of the two servomotors are fully dependent on the mechanism configuration and the screw parameters if the inertial effects are neglected. With the change of the mechanism configuration, the mechanical advantages changes non-linearly.

2.2.2. Kinematic modelling

To derive the kinematical model of the composite actuator, the Assur's Group method [15] is applied. Its basic idea is divide-and-conquer. First, the mechanism is divided into three groups: Group 1 consists of the frame and the input slide A_1 driven by servomotor M_1, Group 2 consists of the frame and the input slide A_2 driven by servomotor M2, and Group 3 is made of Link A_1B_1, Link A_2B_2, Link CD and Link B_1B_2C. Next, the kinematical model for each group is derived. Finally, by combining them together, the kinematical model of the mechanism is founded.

a. Group 1: It is a Class I Assur's group. From Figure 6, it is seen that the position of the joint point A_1 is:

$$\begin{cases} x_1 = d_{1x} + s_1 \cos(\dfrac{3\pi}{2} + \phi_1 + \theta_1) \\ y_1 = d_{1y} + s_1 \sin(\dfrac{3\pi}{2} + \phi_1 + \theta_1) \end{cases} \tag{4}$$

with $s_1 = s_{10} + L_1 \dfrac{\phi_1}{2\pi}$, where s_{10} is the initial displacement and ϕ_1 is the angular displacement of servomotor M_1. d_{1x} and d_{1y} are positions of the original point of slide joint A_1.

b. Group 2: Similar to Group 1, it is a Class I Assur's group and the position of the joint point A_2 is as follows:

$$\begin{cases} x_2 = d_{2x} + s_2 \cos(\frac{\pi}{2} + \phi_2 + \theta_2) \\ y_2 = d_{2y} + s_2 \sin(\frac{\pi}{2} + \phi_2 + \theta_2) \end{cases} \tag{5}$$

with $s_2 = s_{20} + L_2 \frac{\phi_2}{2\pi}$, where s_{20} is the initial displacement and ϕ_2 is the angular displacement of servomotor M_2. d_{2x} and d_{2y} are positions of the original point of slide joint A_2.

c. Group 3: It is an RR-RR-RP-type class-III Assur's group whose solution is numerical instead of analytical. It is modeled as an optimization question where traditional methods can be applied to find the answer. Details can be found in [15].

By single and double differentiations of the displacement formulas with respect to time respectively, the velocity and acceleration can be found as well.

2.2.3. Dimensional conditions and discussions

The mechanism is a parallel mechanism having two DOFs. To make the payload distributed over the mechanical parts nearly uniformly, the mechanism should be symmetrical in configuration and dimension. Thus, we have $\phi_1 = 0°$, $\phi_1 = 180°$, $\delta = 0°$, $\gamma_1 = 0°$, $\gamma_2 = 0°$, $l_1 = l_2$, $l_{31} = l_{32} = l_3 / 2$, $L_1 = L_2 = L$ for the design in Figure 6. The simplified mechanism is depicted in Figure 7.

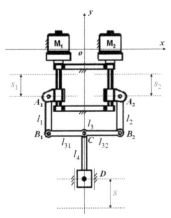

Figure 7. The simplified case for the composite actuator design

For the simplified mechanism, the performance indices of the mechanical advantage expressed in Eq.(1) are simplified as follows:

$$\eta_{ma1} = \frac{L}{2\pi} \cdot \frac{\cos\theta_1 \sin\theta_2}{\cos\delta \sin(\theta_2 - \theta_1)} \tag{6}$$

$$\eta_{ma2} = \frac{L}{2\pi} \cdot \frac{\cos\theta_2 \sin\theta_1}{\cos\delta \sin(\theta_1 - \theta_2)} \tag{7}$$

$$\eta_{ma} = \frac{L}{2\pi} \tag{8}$$

From Eqs.(4)-(8), it is obvious that the driving torques of the two servomotors are fully dependent on the mechanism configuration and the screw parameters if inertial effects are neglected. More importantly, although the driving torques of the two servomotors are time-variant and different, the mechanical advantage of the whole system keeps constant and is dependent on the screw lead merely. Hence, the composite actuator is able to accommodate an un-synchronization between the two servomotors without lowering the mechanical advantage. This is beneficial for the applications of compositie actuators.

2.3. The punching mechanism design

The punching mechanism is one of the most important parts for a mechanical press. One question for punching mechanism design is how to optimize the mechanism with better performances such as reducing torque requirement from servomotors and shortenning working cycle.

2.3.1. The configuration of the punching mechanism

The servo mechanical press developed in this chapter is based on a symmetrical double-knuckle linkage as shown in Figure 5. This punching mechanism has two prismatic (P) joints and eight revolute (R) joints with five independent design parameters, including three link lengths R_1, R_2, and R_3, as well as two assembly distances L and e. The upper P-joint is an active joint actuated by the redundant actuation module detailed above. The bottom P-joint is a passive joint describing the movable connection between the ram and the frame. Therefore, the bias of the ram depends on the manufacturing and assembly accuracies of the links and has no relation to the synchronization of the two servomotors.

2.3.2. Assembly conditions

Half of the punching mechanism is composed of two loops. The upper loop contains Links AB, BC, the active slide C, and the frame. The bottom loop contains Links AB, BD, the ram slide D, and the frame. Considering the upper loop of the punching mechanism, the assembly condition is derived based on triangle inequality as follows:

$$\begin{aligned} R_1 + R_2 > L \quad &\text{(a)} \\ |R_1 - R_2| < L \quad &\text{(b)} \end{aligned} \tag{9}$$

Taking into account the on-site assembly operation, Links AB and BD are assembled firstly in gravitational line, and then Link BC is connected to the two links at Joint B. Besides, Link BD should be connected between Joint B and Joint D at any given configuration of the punching mechanism. Hence, the assembly offset e is set as 0 in this design and Links BC and BD should satisfy following relations

$$R_2 \geq L \qquad \text{(a)}$$
$$R_3 > R_1 \qquad \text{(b)} \tag{10}$$

Besides, the knuckle formed by Links AB and BD is not expected to change folding direction during stamping operation to avoid motion fluctuation of the ram near bottom dead center (BDC). This is possible by controlling the input motion of the active P-joint.

2.3.3. The performance indexing

Generally, a machine design relates to a couple of performance indices. As to the servo mechanical press, the punching mechanism should have a high mechanical advantage to lower the torque required from servomotors. Therefore, the mechanical advantage is an important performance index for the press design. For a press design, the expected tonnage of the press is known. Hence, it is straightforward to use the mechanical advantage reciprocal of the punching mechanism as a performance index that is easily derived according to theoretical mechanics as follows:

$$\eta_{ma} = \frac{F_{in}}{F_{out}} = \frac{\sin\theta_2 \sin(\theta_3 - \theta_1)}{\sin\theta_3 \sin(\theta_2 - \theta_1)} \tag{11}$$

where, F_{in} and F_{out} are input and output forces of the punching mechanism respectively, and θ_1, θ_2 as well as θ_3 are angles measured from positive horizontal axis to Links AB, CB and BD respectively. Obviously, θ_1, θ_2 and θ_3 are functions of input motion of the active P-joint.

Also the force distribution among the links is important for equal life cycle structure design of the parts of the servo press. Since the punching force is originated from the plastic deformation of the workpieces during stamping operation, the force distribution among the links can be measured in terms of ratio of the link force over the output force as follows:

$$\eta_1 = \frac{\sin(\theta_2 - \theta_3)}{\sin\theta_3 \sin(\theta_2 - \theta_1)} \qquad \text{(a)}$$

$$\eta_2 = \frac{\sin(\theta_1 - \theta_3)}{\sin\theta_3 \sin(\theta_2 - \theta_1)} \qquad \text{(b)} \tag{12}$$

$$\eta_3 = \frac{1}{\sin\theta_3} \qquad \text{(c)}$$

where, η_1, η_2 and η_3 are related to Links AB, BC and BD respectively.

2.3.4. The dimension design with visual global optimization

The dimension synthesis is very important for mechanism design. Current researches focus on how to build optimized design models. These models are solved by optimization methods that lead to one global/local optimized solution. Apparently, the solution depends on given conditions and/or given weighting factors for multi-goal design problems. A slim variation of the weighting factors often leads to a big change of the solution that confuses the designers. To make multi-goal optimization more straightforward, a special method was proposed by Yang, Gao and their followers that transforms the unbounded dimension space into a limited solution space [16,17], where performance atlases were displayed to show a global image of performance indices. Over these atlases, the area with global optimization is intuitive.

a. The bounded feasible solution space

The punching mechanism of the servo mechanical press studied in this paper is a symmetrical double- knuckle linkage with four independent design parameters, R_1, R_2, R_3 and L as discussed above. Theoretically, the links can have their lengths up to infinity that makes the solution space an unbounded one. According to [16, 17], the link lengths can be normalized to form a limited solution space.

Let

$$r_i = \frac{R_i}{R}, \quad i = 1,2,3 \qquad \text{(a)}$$

$$l = \frac{L}{R} \qquad \text{(b)} \qquad\qquad (13)$$

where, $R = \dfrac{R_1 + R_2 + R_3 + L}{4}$, r_i (i=1,2,3) and l are normalized/non-dimensional parameters, and R is an average value.

Therefore, one has

$$r_1 + r_2 + r_3 = 4 - l \qquad\qquad (14)$$

Considering the assembly conditions discussed above, the normalized parameters should satisfy following relations:

$$
\begin{array}{ll}
r_1 + r_2 > l & \text{(a)} \\
|r_1 - r_2| < l & \text{(b)} \\
r_2 \geq l & \text{(c)} \\
r_3 > r_1 & \text{(d)}
\end{array}
\qquad (15)
$$

Eq.(14) and inequality (15) construct the mathematic model of the bounded solution space for the servo mechanical press design. Hence, a polygonal solution slice with given L can be taken out from the solution polyhedron represented by this model as shown in Figure 8. In this figure, the feasible solution locates at the shaded triangle area surrounded by the lines of $r_2 - r_1 = l$, $r_2 = l$ and $r_3 = r_1$. Obviously, $0 < r_1 \le 2 - l$, $l \le r_2 \le (4 + l)/3$ and $(4 - 2l)/3 \le r_3 \le 4 - 2l$. The feasible solution space for the servo press design is limited by all the possible design constraints.

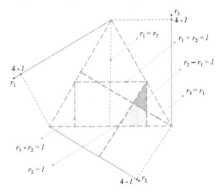

Figure 8. A slice of the bounded solution space for the servo press

b.   The performance atlases

Using Eqs. (11)-(13), the performance atlases can be drawn over the bounded feasible solution space as shown in Figure 9 with respect to r_1, r_2 and r_3 at given l. Here, the maximum value of the performance indices is displayed as example. From Figure 9, it is straightforward to see the variation trend of the performance indices with respect to the deimensions. For example, it is easy to see that the force load of Link AB increases and the mechanical advantage decreases while r_2 increases. Obviously, the performance atlases exhibit the global distribution of the perfermance indices that are helpful to the dimension synthesis leading to near global optimization.

Similar to above, any other interested performance index can be expressed as performance atlas related to link dimensions.

c.   The visual dimension design

With the help of several performance atlases, the domain containing satisfactory solutions can be determined directly to accommodate several design goals. Then a good solution can be selected from the domain based on expertises, experiences or other rules. Also an optimized solution is able to be achieved further by optimization algorithms starting from points at the selected domain that ensures global optimization.

2.3.5. Case study

To calculate the performances of different designs of servo presses, the well-known deep drawing process is introduced to calculate the performance indices. The main performance parameters of the press are assumed as follows: (1) maximum slide stroke: 200mm; (2) travel of nominal capacity: 8mm; and (3) stroke number: 40rpm.

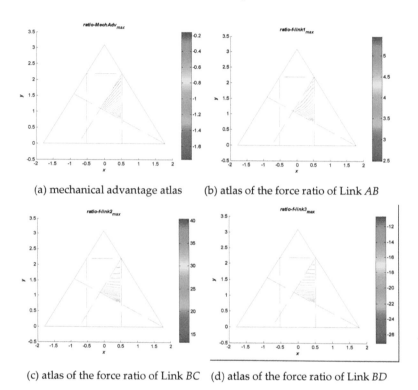

(a) mechanical advantage atlas   (b) atlas of the force ratio of Link AB

(c) atlas of the force ratio of Link BC (d) atlas of the force ratio of Link BD

Figure 9. Performance atlases

	r_1	r_2	r_3	L
1	0.4	0.8	2.0	0.8
2	0.4	1.2	1.4	1.0
3	0.6	1.1	1.3	1.0
4	0.8	1.0	1.2	1.0
5	0.1	1.3	1.4	1.2

Table 1. The five design cases

Here are five cases listed in Table 1 as example. To calculate the performance indices, the average length R can be set according to the maximum slide stroke 200mm. In Figure 10, the maximum mechanical advantage reciprocals within the travel of nominal capacity 8mm for the five cases are displayed over the bounded feasible solution space. The dimension points for the middle three cases locate at one same solution slice. It can be seen that the value range grows bigger with the decreasing of L. From the graphs in Figure 10, the five designs can be sorted in terms of the maximum mechanical advantage reciprocals as 1, 4, 2, 3 and 5. And it is clear that to what extent the given designs approach the global optimization in terms of one performance index by observing the corresponding performance atlas. The maximum mechanical advantage reciprocals are listed in Table 2 for the five cases.

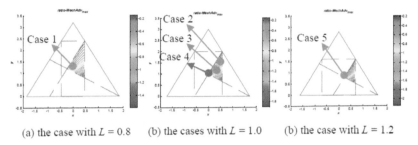

(a) the case with $L = 0.8$ (b) the cases with $L = 1.0$ (b) the case with $L = 1.2$

Figure 10. The designs in feasible solution space

	Case 1	Case 2	Case 3	Case 4	Case 5
η_{ma}	1.8057	4.0148	7.242	2.522	13.4863
H_1	0.8057	3.0148	8.3317	1.522	12.4863
H_2	2.0618	4.1998	7.8418	3.3132	13.5197
η_3	1.0138	1.5874	2.0895	2.6967	1.3797

Table 2. The maximum performance comparisons

3. The design of programmable stamping operation

The most outstanding advantage of the servo press is the flexibility, i.e. the programmability of its ram motions to accommodate different stamping operations such as drawing, stamping, blanking, coining, etc. To design the stamping operation, a pseudo-NURBS method is presented to model the ram motion where the displacement and velocity demands of the stamping operation (ram motion) are transformed to weight constrains, and inequalities are formed by using weights as variables. An optimized ram motion can be generated to match the deformation of work piece material that avoids large transient force and vibration by smooth approaching and slow releasing of the ram, and improves stamping productivity via quick return of the ram.

3.1. The pseudo-NURBS representation of stamping operation

The ram of the press only moves along a line segment within a stroke reciprocatedly and the ram motion is changeable in terms of time. To design the ram motion, a reasonable mathematical expression should be determined firstly. As a well-known tool in computer graphics and CAGD [18], the NURBS offers a common mathematical representation for free-form surfaces/curves and commonly used analytical shapes such as natural quadrics, torii, extruded surfaces and surfaces of revolution. It is easy and straightforward to change the shape through the manipulation of control points, weights and knots. And degree elevation, splitting, knot insertion and deletion and knot refinement offer a wide range of tools to design and analyze shape information. In this chapter, a pseudo-NURBS method is proposed to express the ram motion by introducing time parameter into the NURBS representation. A NURBS curve is a defined by the following equation:

$$c(u) = \sum_{i=0}^{n} P_i R_{i,k}(u), \qquad R_{i,k}(u) = \frac{\omega_i N_{i,k}(u)}{\sum_{i=0}^{n} \omega_i N_{i,k}(u)} \tag{16}$$

where, P_i are vectors composed of x and y coordinates of the control points, ω_i are positive real weights for each control point, and point of $c(u)$ is called curve point. $R_{i,k}(u)$ are NURBS basis functions, and $N_{i,k}(u)$ are B-spline basis functions of degree k-1. The B-spline basis function $N_{i,k}(u)$ is expressed by de Boor-Cox [8] as follows:

$$N_{i,0}(u) = \begin{cases} 1, u_i \leq u \leq u_{i+1} \\ 0, otherwise \end{cases} \tag{17}$$

$$N_{i,k}(u) = \frac{u - u_i}{u_{i+k} - u_i} N_{i,k-1}(u) + \frac{u_{i+1} - u}{u_{i+k+1} - u_{i+1}} N_{i+1,k-1}(u) \tag{18}$$

where, the interval $[u_i, u_{i+1}]$ is called the ith knot span, and $U=\{u_0, u_1, \ldots u_m\}$ ($m=n+k+1$) is the knot vector. A knot vector is non-periodic if the first and last knots are repeated with multiplicity k+1, i.e. $U=\{u_0, \ldots, u_0, u_{k+1}, \ldots, u_{m-k-1}, u_m, \ldots u_m\}$. For most practical applications, $u_0 = 0$ and $u_m = 1$. Knots $u_{k+1}, \ldots, u_{m-k-1}$ are called interior. For planar curve, $U = \{0, 0, 0, u_{k+1}, \ldots u_{m-k-1}, 1, 1, 1\}$.

The most significant properties of the NURBS curve are:

- Locality: A movement of the control point P_i affects the curve only in the interval $[u_i, u_{i+k+1}]$.
- Affine and projective invariance: A NURBS curve is affinely and projectively invariant, i.e. transforming the curve by an affine or perspective transform is equivalent to transforming the control points.
- Differentiability: Same differentiability properties as the basis functions.

Different from the geometric shape, the time parameter t should be introduced in order to express the ram motion. To apply NURBS expression for the ram motion, following issues are considered:

1. time parameter t: To build the relationship of the ram displacement with respect to time, the stamping time and displacement are normalized by dividing by stampling cycle and stroke respectively.
2. control points and curve points: As shown in Figure 11, the control points P_i should be determined according to Eq. (16) given the curve point Q_i.

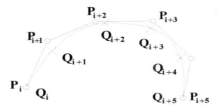

Figure 11. Control points and curve points

3.   boundary constraints: The velocity and acceleration at the firstpoint and the endpoint should be equal to zero for the stamping cycle.

4.   knot insertion. The chord method is applied as follows:

$$d = \sum_{r=1}^{n} |Q_r - Q_{r-1}| \tag{19}$$

$$t_0 = 0, \quad t_r = t_{r-1} + \frac{|Q_r - Q_{r-1}|}{d}, \quad t_n = 1 \tag{20}$$

where, Q_r is curve point, n is the number of the curve points, and t_r is intermittent element of knot vector. The knot is expressed as

$$u_{j+p} = \frac{1}{p} \sum_{i=j}^{j+p-1} t_i \quad j = 1, 2, \ldots, n-p \tag{21}$$

Hence, a pseudo-NURBS expression is proposed for the ram motion modeling. To assure the velocities and accelerations of the ram at the first points and endpoints zero, following constraints should be followed:

$$C^{(i)}(t_0) = D_s^{(i)}, \quad C^{(j)}(t_n) = D_e^{(j)}, \ldots i = 1,2,\ldots,sp, j = 1,2,\ldots,ep \tag{22}$$

where, $C^{(i)}(t_0)$ refers to i-th derivative at the first point, $C^{(j)}(t_n)$ means j-th derivative at the endpoint. Equations with the sum number of $sp+ep+1$ are formed from Eqs.(16) and (22) and the control points of the same number need to be determined.

$$C^{(j)}(x) = \sum_{i=0}^{n} R_{i,k}^{(j)}(x) P_i \tag{23}$$

$$R_{j,k}^{(1)}(x) = \omega_j N_{j,k}^{(1)}(x) / \sum_{i=0}^{n} \omega_i N_{i,k}(x) - \omega_i N_{j,k}(x) \sum_{i=0}^{n} \omega_i N_{i,k}^{(1)}(x) / \left[\sum_{i=0}^{n} \omega_i N_{i,k}(x) \right]^2 \tag{24}$$

$$R_{j,k}^{(2)}(x) = \omega_j N_{j,k}^{(2)}(x) / \sum_{i=0}^{n} \omega_i N_{i,k}(x) - \left[2\omega_j N_{j,k}^{(1)}(x) \sum_{i=0}^{n} \omega_i N_{i,k}^{(1)}(x) + \omega_j N_{j,k}(x) \sum_{i=0}^{n} \omega_i N_{i,k}^{(2)}(x) \right]$$

$$/\left[\sum_{i=0}^{n}\omega_i N_{i,k}(x)\right]^2 + 2\omega_j N_{j,k}(x)\left[\sum_{i=0}^{n}\omega_i N_{i,k}^{(1)}(x)\right]^2 / \left[\sum_{i=0}^{n}\omega_i N_{i,k}(x)\right]^3 \tag{25}$$

$$N_{i,p}^{(j)}(x) = p\left(\frac{N_{i,p-1}^{(j-1)}(x)}{u_{i+p}-u_i} - \frac{N_{i+1,p-1}^{(j-1)}(x)}{u_{i+p-1}-u_{i+1}}\right) \tag{26}$$

Combining Eqs.(16), and (22) through (26), following equation can be got:

$$
\begin{bmatrix}
1 \\
R_0^{(1)} & R_1^{(1)} \\
R_0^{(2)} & R_1^{(2)} & R_2^{(2)} \\
& R_1 & R_2 & R_3 & R_4 \\
& & R_2 & R_3 & R_4 & R_5 \\
& & & R_3 & R_4 & R_5 & R_6 \\
& & & & R_5 & R_6 & R_7 & R_8 \\
& & & & R_5 & R_6 & R_7 & R_8 \\
& & & & & R_7 & R_8 & R_9 & R_{10} \\
& & & & & & R_8 & R_9 & R_{10} & R_{11} \\
& & & & & & & R_{10}^{(2)} & R_{11}^{(2)} & R_{12}^{(2)} \\
& & & & & & & & R_{11}^{(1)} & R_{12}^{(1)} \\
& & & & & & & & & 1
\end{bmatrix}
\begin{bmatrix}
P_0 \\ P_1 \\ P_2 \\ P_3 \\ P_4 \\ P_5 \\ P_6 \\ P_7 \\ P_8 \\ P_9 \\ P_{10} \\ P_{11} \\ P_{12}
\end{bmatrix}
=
\begin{bmatrix}
Q_0 \\ D_s^{(1)} \\ D_s^{(2)} \\ Q_1 \\ Q_2 \\ Q_3 \\ Q_4 \\ Q_5 \\ Q_6 \\ Q_7 \\ D_e^{(2)} \\ D_e^{(1)} \\ Q_8
\end{bmatrix}
\tag{27}
$$

By solving Eq.(27), the control points P_0, P_1..., and P_{12} are determined and therefore the ram motion can be solved according to Eq.(16). Thanks to the locality of NURBS, the ram motion can be refined piecewisely through the manipulation of corresponding control points, weights and knots.

3.2. The influence of weights to the fluctuations of ram velocity and acceleration

To check the influence of weights to the ram motions, let $[\omega_1]$ = [1, 1, 1, 1, 1, 1, 1, 1, 1, 1, 1, 1, 1], $[\omega_2]$ = [1, 0.5, 0.6, 0.8, 1,0.9, 0.8, 0.4, 0.8, 1.2, 0.5, 0.8, 1], and $[\omega_3]$ = [1, 1, 1, 1, 0.5, 0.6, 0.8, 0.1, 0.8, 1, 1, 1, 1]. The corresponding ram motions are dicpited in Figure 12 resepctively. It can be found that weight ω_i (i=8) should be adjusted that can lower the fluctuation of the ram velocity and acceleration and decrease the shock, vibration and noise during stamping operation, and decrease the peak power demand of the servomotors.

3.3. The influence of weights and knot vector to the fluctuations of ram velocity and acceleration

To check the influence of weights and knot vector to the ram motions, let $[\omega_1]$ = [1, 1, 1, 1, 1, 1, 1, 1, 1, 1, 1, 1, 1], $[\omega_2]$ = [1, 0.5, 0.6, 0.8, 1, 0.9, 0.8, 0.4, 0.8, 1.2, 0.5, 0.8, 1], $[\omega_3]$ = [1, 1, 1, 1, 0.5,

0.6, 0.8, 0.05, 0.8, 1.2, 0.5, 0.8, 1], and $U_2 = [0, 0, 0, 0, 0.0476, 0.1111, 0.254, 0.4444, 0.619, 0.7619,$
0.8571, 0.92, 0.995, 1, 1, 1, 1]. The corresponding ram motions are dicpited in Figure 13
resepctively. It can be found that the peak velocity and acceleration of the return period can
be lowered up to 20%-30% and the ram velocity lowered 40% when the upper die touches
the workpiece by adjusting the knots corresponding to quick return period.

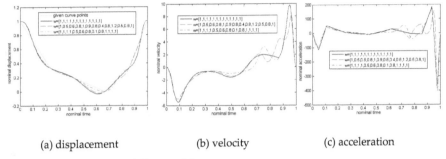

(a) displacement   (b) velocity   (c) acceleration

Figure 12. Ram motion under different weights

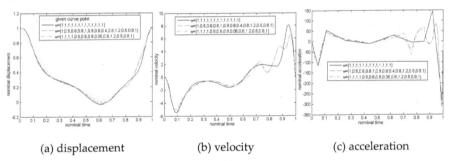

(a) displacement   (b) velocity   (c) acceleration

Figure 13. Ram motion under different weights and knot vectors

3.4. The self-adaption of the weights due to optimization of the ram motioin

Under the same group of demands over the displacement, velocity of the ram motion, a
large number of NURBS curves can be achieved and hence the optimized one is able to
select by optimization.

3.4.1. The self-adaption due to the ram displacement

Due to the local support of the NURBS expression, the ram motion can be discretized into
several segments to set up the constraint inequalities. The m curve segments are denoted as
$l_1, l_2, \ldots,$ and l_m. Supposed displacement demands over curve segment $l_j, s_1 \leq l_j \leq s_2$, i.e.

$s_1 \leq \dfrac{\sum_{i=0}^{n} d_i \omega_i N_{i,k}(u)}{\sum_{i=0}^{n} \omega_i N_{i,k}(u)} \leq s_2$, where knot vector $u \in [u_{j1}, u_{j2}]$. To satisfy this in-equality demand, let the two endpoints, maximum and minimum values over l_j locate at the domain formed by s_1 and s_2, i.e.

$$s_1 \leq \frac{\sum_{i=0}^{n} d_i \omega_i N_{i,k}(u_{j1})}{\sum_{i=0}^{n} \omega_i N_{i,k}(u_{j1})} \leq s_2 \tag{28}$$

$$s_1 \leq \frac{\sum_{i=0}^{n} d_i \omega_i N_{i,k}(u_{j2})}{\sum_{i=0}^{n} \omega_i N_{i,k}(u_{j2})} \leq s_2 \tag{29}$$

$$s_1 \leq \max \left(\frac{\sum_{i=0}^{n} d_i \omega_i N_{i,k}(u)}{\sum_{i=0}^{n} \omega_i N_{i,k}(u)} \right) \leq s_2 \tag{30}$$

$$s_1 \leq \min \left(\frac{\sum_{i=0}^{n} d_i \omega_i N_{i,k}(u)}{\sum_{i=0}^{n} \omega_i N_{i,k}(u)} \right) \leq s_2 \tag{31}$$

3.4.2. The self-adjustment due to velocity of the ram

Supposed velocity demand over curve section l_j $v_1 \leq v_{ij} \leq v_2$, where $u \in [u_{j1}, u_{j2}]$. For $v_{ij} = \sum_{i=0}^{n} R'_{i,k}(u)d_i$, we have $v_1 \leq \sum_{i=0}^{n} R'_{i,k}(u)d_i \leq v_2$, i.e.

$$v_1 \leq \sum_{i=0}^{n} \left(\frac{\frac{\omega_i N'_{i,k}(u)}{\sum_{j=0}^{n} \omega_j N_{j,k}(u)} - \omega_i N_{i,k}(u)\sum_{j=0}^{n} \omega_j N'_{j,k}(u)}{[\sum_{j=0}^{n} \omega_j N_{j,k}(u)]^2} \right) d_i \leq v_2 \tag{32}$$

Let $p_1 = [\sum_{j=0}^{n} \omega_j N_{j,k}(u)]^2$, $p_2 = \sum_{i=0}^{n} \left(\frac{\omega_i N'_{i,k}(u)}{\sum_{j=0}^{n} \omega_j N_{j,k}(u)} \right) d_i$, and $p_3 = \sum_{i=0}^{n} \left(\omega_i N_{i,k}(x) \sum_{j=0}^{n} \omega_j N'_{j,k}(x) \right) d_i$. Eq.(32) is transformed as

$$v_1 \leq \frac{p_2 - p_3}{p_1} \leq v_2 \tag{33}$$

To satisfy Eq.(33), make sure the values of the first point and endpoint, maximum and minimum values over v_{ij} locating within the domain spaned by v_1 and v_2, i.e.

$$v_1 \leq \frac{p_2(u_{j1}) - p_3(u_{j1})}{p_1(u_{j1})} \leq v_2 \tag{34}$$

$$v_1 \leq \frac{p_2(u_{j2}) - p_3(u_{j2})}{p_1(u_{j2})} \leq v_2 \tag{35}$$

$$v_1 \leq \max\left(\frac{p_2(u)-p_3(u)}{p_1(u)}\right) \leq v_2 \qquad (36)$$

$$v_1 \leq \min\left(\frac{p_2(u)-p_3(u)}{p_1(u)}\right) \leq v_2 \qquad (37)$$

3.4.3. The solution of the optimized weights

To lower the velocity fluctuation and ease the shock, vibration and noise to improve the workpiece's stamping quality, goal function can be defined by controlling the velocity of the ram as follows:

$$F(\omega) = \min\left(\Sigma_{i=0}^{n} \mid v_i(\omega_i)\right) \qquad (38)$$

where, $v_i(\omega_i) = \Sigma_{i=0}^{n} R_{i,k}'(u)d_i$.

Considering some of the weights have no relation to the stamping operation, these weights are prescribed and other weights are selected to be optimized under the demands of displacement, velocity and above goal function by iterative solution.

3.5. Case study

3.5.1. The problem definition

Take the deep drawing operation as example. The performance demands include: (1) uniform stamping operation: nearly constant stamping velocity (100-300mm/s) for drawing period, (2) quick return: high ram speed (330-500mm/s) for return period, and (3) small fluctuation of ram velocity and acceleration during drawing and releasing periods.

3.5.2. The optimization

To express the relationship between the ram motion and time by NURBS, the ram displacement and time are normalized firstly as follows:

t_0 / t =[0, 0.67/4, 0.99/4, 1.80/4, 2.25/4, 2.97/4, 3.37/4, 3.69/4, 1], and

s_0 / s =[1, 186.78/400,106.67/400,44.44/400,0,62.22/400,177.78/400,275.56/400,1].

Correspondingly, the velocity constraints are normalized to be nominal velocity v_0 expressed as $v_0 = \frac{t}{s}\frac{ds_0}{dt_0}$. The real ram velocity can be derived as $v = \frac{s}{t}v_0$.

Because weights ω_i only affects the curve shape over the knot span $[u_i, u_{i+k+1}] \in [u_k, u_{n+1}]$, here weights $\omega_3, \omega_4, \omega_5, \omega_6$ are selected as optimized variables that relate to the drawing and releasing periods. The total weights are set as ω= [0.5, 1, ω_3, ω_4, ω_5, ω_6, 0.5, 1, 0.8, 0.5, 0.8, 0.7, 0.9].

To assure the continousness of ram displacement, velocity and acceleration, NURBS of $k = 3$ is applied here. The constraints relating the punching speed are listed as:

1.   drawing period: -300mm/s < $v_1(t)$ < -100mm/s, and 0.99s < t <2.25s, and
2.   releasing period: 330mm/s < $v_2(t)$ < 500mm/s, and 2.51s < t < 2.97s.

To limit the velocity fluctuation, the goal function is defined as $F(\omega) = \min \left(\Sigma_{i=0}^{n} \mid v_i \mid \right)$.

Hence, we can get the optimized weights ω_3 =0.5, ω_4 =0.7, ω_5 =0.8, and ω_6 =0.7.

3.5.3. Performance analysis

As discussed above, the optimized weights are

ω_a = [0.5, 1, 0.5, 0.7, 0.8, 0.7, 0.5, 1, 0.8, 0.5, 0.8, 0.7, 0.9].

Here another two weights are set for comparison as:

ω_b = [0.5, 1, 0.9, 0.4, 0.5, 0.2, 0.5, 1, 0.8, 0.5, 0.8, 0.7, 0.9]
ω_c = [0.5, 1, 0.3, 0.5, 0.1, 1, 0.5, 1, 0.8, 0.5, 0.8, 0.7, 0.9]

Figure 14 displays the ram motion curves under weights ω_a, ω_b, and ω_c respectively. From the figures, it can be seen that the ram motion of ω_a is the nearest to the desired trajectory for drawing and releaseing period, the ram velocity curve of ω_a is more flat for drawing period, and the fluctuation of the ram acceleration of ω_a is the least that helps to decrease shock, vibration and noise. This is consistent with the comparison of goal functions. The goal function values are 20441, 20457 and 20606 under ω_a, ω_b and ω_c respectively.

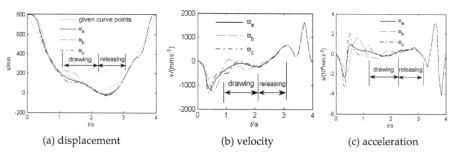

(a) displacement   (b) velocity   (c) acceleration

Figure 14. The optimized ram motion

3.5.4. The prototype development

To validate the new concept, a servo press prototype with 200 ton punching force has been developed in our lab. As shown in Figure 15, the machine is more than 5 meters high and 2 meters wide. Experimental studies are being performed on this prototype including dynamic performance, dynamic control and stamping process planning.

Figure 15. The servo press prototype with 2-servomotor redundant actuation of 200 tonnage (photo)

4. Conclusion

In this chapter, a new concept of programmable metal forming press of redundant parallel actuation is presented to make it possible to develop a larger servo mechanical press using servomotors available. Based on the above discussions, following conclusions can be drawn:

1. A servo mechanical press is different from a traditional mechanical press in such a way that it involves servo actuation design with programmable motors and punching mechanism design with higher mechanical advantage.
2. The tonnage of servo mechanical presses is limited by the capacity of servomotors available to some extent. The new servo mechanical press with redundant actuation proposed in this chapter is a feasible option to develop a larger machine based on servomotors available. This redundant actuation scheme is able to accommodate an un-synchronization between the two servomotors.
3. The mechanism design for the redundant actuation should be considered carefully that maintains the performance of the redundant actuation.
4. The ram motion of the servo press can be optimized that requires a tool to express. The pseudo-NURBS representation proposed is effective to model the ram motion and form an optimization problem. Simulation shows that the optimized ram motion improves stamping operation with higher productivity and avoids large transient force and vibration by smooth approaching and slow releasing of the ram.
5. A computer programme based on the pseudo-NURBS method is developed in the laboratory embedded in the servo press that facilitates the press operation.

Author details

Weizhong Guo and Feng Gao
State Key Lab of Mechanical System and Vibration, Shanghai Jiao Tong University, Shanghai, China

Acknowledgement

The author thanks the partial financial supports under the projects from the National Natural Science Foundation of China (NSFC Grant No. 50875161), Program for New

Century Excellent Talents in University (Grant No. NCET-10-0567), the Research Fund of State Key Lab of MSV, China (Grant No.MSV-ZD-2010-02), and National Science and Technology Major Projecct, China (Grant No. 2010ZX04004-112).

5. References

[1] http://www.komatsusanki.co.jp
[2] http://www.amada.com
[3] http://www.amino.co.jp/index.html
[4] http://www.chinfong.com.cn/english/index.htm
[5] Shivpuri, R.S. Yossifon, A servo motor driven multi-action press for sheet metal forming, International Journal of Machinery Tool Manufacture, 1991, 31: 345-359
[6] Yossifon.S, Shivpuri.R. (1993) Design considerations for the electric servo-motor driven 30 ton double knuckle press for precision forming. International Journal of Machinery Tool Manufacture, 33(2): 209-222.
[7] Yan, H. S., Chen, W. R. (2000) On the Output Motion Characteristics of Variable Input Speed Servo-Controlled Slider-Crank Mechanisms, Mechanism and Machine Theory, 35: 541-561.
[8] R. Du, W. Z. Guo (2003) The Design of a New Metal Forming Press With Controllable Mechanism [J]. ASME Journal of Mechanical Design, 125(3): 582-592.
[9] Guo, W. Z., Du, R. (2005) A New Type of Controllable Mechanical Press-Motion Control and Experimental Validation, ASME Journal of Manufacturing Science and Engineering, 127(4): 731-742.
[10] W.Z. Guo, F. Gao, R. Du (2008) The Design and Prototype of a Servo Mechanical Press with Hybrid Inputs, Proceedings of the 8th International Conference on Frontiers of Design and Manufacturing, Sept. 23-26, 2008, Tianjin, China
[11] Tokuz, L. C. and Jones, J. R. (1991) Programmable Modulation of Motion Using Hybrid Machines, Proceedings of IMECHE, C413/071, pp. 85-91.
[12] Tokuz, L. C., 1992, Hybrid Machine Modeling and Control, Ph.D. Dissertation, Liverpool Polytechnic University.
[13] GUO Wei-zhong, GAO Feng (2009) Design of a Servo Mechanical Press with Redundant Actuation. Chinese Journal of Mechanical Engineering, 22(4): 574-579.
[14] Yongjun Bai, Feng Gao, Weizhong Guo, Yi Yue (2011) Study of the dual screw actuation for servo mechanical presses, Proceedings of 2nd IFToMM International Symposium on Robotics and Mechatronics, Shanghai, China, November 3-5, 2011
[15] Liang, C. G. and Ruan, P. S. (1986) Computer-Aided Design of Linkages, Beijing: Machine Press (in Chinese)
[16] Yang Ji-hou, Gao Feng. Solution Space and Performance Atlases of the Four-bar Mechanism [M], Beijing: China Machine Press, 1989 (in Chinese)
[17] Gao F., Zhang X.Q., Zhao Y.S., Wang H.R. (1996) A physical model of the solution space and the atlases of the reachable workspaces for 2-DOF parallel planar manipulators, Mechanism and Machine Theory, 31(2):173-184
[18] Les Piegl (1991) On NURBS: a survey, IEEE Computer Graphics & Applications, (1): 55-71.

Permissions

The contributors of this book come from diverse backgrounds, making this book a truly international effort. This book will bring forth new frontiers with its revolutionizing research information and detailed analysis of the nascent developments around the world.

We would like to thank Mohsen Kazeminezhad, for lending his expertise to make the book truly unique. He has played a crucial role in the development of this book. Without his invaluable contribution this book wouldn't have been possible. He has made vital efforts to compile up to date information on the varied aspects of this subject to make this book a valuable addition to the collection of many professionals and students.

This book was conceptualized with the vision of imparting up-to-date information and advanced data in this field. To ensure the same, a matchless editorial board was set up. Every individual on the board went through rigorous rounds of assessment to prove their worth. After which they invested a large part of their time researching and compiling the most relevant data for our readers. Conferences and sessions were held from time to time between the editorial board and the contributing authors to present the data in the most comprehensible form. The editorial team has worked tirelessly to provide valuable and valid information to help people across the globe.

Every chapter published in this book has been scrutinized by our experts. Their significance has been extensively debated. The topics covered herein carry significant findings which will fuel the growth of the discipline. They may even be implemented as practical applications or may be referred to as a beginning point for another development. Chapters in this book were first published by InTech; hereby published with permission under the Creative Commons Attribution License or equivalent.

The editorial board has been involved in producing this book since its inception. They have spent rigorous hours researching and exploring the diverse topics which have resulted in the successful publishing of this book. They have passed on their knowledge of decades through this book. To expedite this challenging task, the publisher supported the team at every step. A small team of assistant editors was also appointed to further simplify the editing procedure and attain best results for the readers.

Our editorial team has been hand-picked from every corner of the world. Their multi-ethnicity adds dynamic inputs to the discussions which result in innovative outcomes. These outcomes are then further discussed with the researchers and contributors who give their valuable feedback and opinion regarding the same. The feedback is then collaborated with the researches and they are edited in a comprehensive manner to aid the understanding of the subject.

Apart from the editorial board, the designing team has also invested a significant amount of their time in understanding the subject and creating the most relevant covers. They scrutinized every image to scout for the most suitable representation of the subject and create an appropriate cover for the book.

The publishing team has been involved in this book since its early stages. They were actively engaged in every process, be it collecting the data, connecting with the contributors or procuring relevant information. The team has been an ardent support to the editorial, designing and production team. Their endless efforts to recruit the best for this project, has resulted in the accomplishment of this book. They are a veteran in the field of academics and their pool of knowledge is as vast as their experience in printing. Their expertise and guidance has proved useful at every step. Their uncompromising quality standards have made this book an exceptional effort. Their encouragement from time to time has been an inspiration for everyone.

The publisher and the editorial board hope that this book will prove to be a valuable piece of knowledge for researchers, students, practitioners and scholars across the globe.

List of Contributors

A. El Hami
LMR, INSA de Rouen, St Etienne de Rouvray, France

B. Radi
LMMI, FST Settat, Settat, Morocco

A. Cherouat
GAMMA3, UTT, Troyes, France

Xin-Yun Wang, Jun-song Jin, Lei Deng and Qiu Zheng
State Key Laboratory of Materials Processing and Die & Mould Technology, Huazhong
University of Science and Technology, Wuhan, China

M. Bakhshi-Jooybari, A. Gorji and M. Elyasi
Faculty of Mechanical Engineering, Babol University of Technology, Babol, Mazandaran,
Iran

Bernd Engel and Johannes Buhl
University of Siegen, Chair of Forming Technology, Siegen, Germany

Marta Oliveira
INEGI – Instituto de Engenharia Mecânica e Gestão Industrial, Portugal

Tetsuhide Shimizu, Ming Yang and Ken-ichi Manabe
Tokyo Metropolitan University, Japan

Javier W. Signorelli and María de los Angeles Bertinetti
Instituto de Física Rosario (IFIR), CONICET–UNR, Rosario, Argentina

Weizhong Guo and Feng Gao
State Key Lab of Mechanical System and Vibration, Shanghai Jiao Tong University,
Shanghai, China

Printed in the USA
CPSIA information can be obtained
at www.ICGtesting.com
JSHW011811301024
72690JS00002B/40